Intelligent Systems Reference Library

Volume 220

Series Editors

Janusz Kacprzyk, Polish Academy of Sciences, Warsaw, Poland

Lakhmi C. Jain, KES International, Shoreham-by-Sea, UK

The aim of this series is to publish a Reference Library, including novel advances and developments in all aspects of Intelligent Systems in an easily accessible and well structured form. The series includes reference works, handbooks, compendia, textbooks, well-structured monographs, dictionaries, and encyclopedias. It contains well integrated knowledge and current information in the field of Intelligent Systems. The series covers the theory, applications, and design methods of Intelligent Systems. Virtually all disciplines such as engineering, computer science, avionics, business, e-commerce, environment, healthcare, physics and life science are included. The list of topics spans all the areas of modern intelligent systems such as: Ambient intelligence, Computational intelligence, Social intelligence, Computational neuroscience, Artificial life, Virtual society, Cognitive systems, DNA and immunity-based systems, e-Learning and teaching, Human-centred computing and Machine ethics, Intelligent control, Intelligent data analysis, Knowledge-based paradigms, Knowledge management, Intelligent agents, Intelligent decision making, Intelligent network security, Interactive entertainment, Learning paradigms, Recommender systems, Robotics and Mechatronics including human-machine teaming, Self-organizing and adaptive systems, Soft computing including Neural systems, Fuzzy systems, Evolutionary computing and the Fusion of these paradigms, Perception and Vision, Web intelligence and Multimedia.

Indexed by SCOPUS, DBLP, zbMATH, SCImago.

All books published in the series are submitted for consideration in Web of Science.

More information about this series at https://link.springer.com/bookseries/8578

Marian Cristian Mihăescu

Editor

Data Analytics in e-Learning: Approaches and Applications

 Springer

Editor
Marian Cristian Mihăescu
University of Craiova
Craiova, Romania

ISSN 1868-4394   ISSN 1868-4408   (electronic)
Intelligent Systems Reference Library
ISBN 978-3-030-96646-1   ISBN 978-3-030-96644-7   (eBook)
https://doi.org/10.1007/978-3-030-96644-7

This Springer imprint is published by the registered company Springer Nature Switzerland AG
The registered company address is: Gewerbestrasse 11, 6330 Cham, Switzerland

Preface

Research efforts in EDM[1] and LAK[2]—these are among the most well-known conferences related to data analytics in e-learning—are currently focusing on usage and integration of machine learning (ML) algorithms for solving various educational problems like knowledge tracing, student performance, and lately computational linguistic. Of course, there are many success stories, mainly due to many software packages and libraries that implement a wide range of ML algorithms. The reasonable ease of usage and integration of ML algorithms in virtually any application opened the way towards enhancing e-learning platforms. Still, several issues need to be well understood by researchers and practitioners: (1) building complex workflows, (2) validating the results, and (3) building new algorithms that solve specific problems. This book may be regarded as an example of passing through all these steps such that the reader may further tackle similar problems.

Nowadays, data analytics—sometimes called data science—represent one of the most common today's buzz words. As there are many quite similar definitions, I will mention the essential ingredients: examining datasets, finding trends, and drawing conclusions. Of course, a description based on these ingredients is too general. Under these circumstances, one aspect worth investigating is how this available setup may transform when used in a specific application domain or even within a particular software application from that domain.

This book represents the digest of the research experience from about 6 years of using data analytics in the practical context of an e-learning application. The book's primary purpose is to provide the reader with a roadmap of key ingredients that may be needed to delve into data analytics for e-learning. Therefore, the book's critical elements in the attempt to showcase a complete workflow are: (1) public dataset, (2) data analysis workflow, (3) interpretable models, (4) enhancing models, (5) increasing engagement, (6) usability evaluation, and (7) building new algorithms.

---

[1] Educational Data Mining, https://educationaldatamining.org/.

[2] Learning Analytics and Knowledge, https://www.solaresearch.org/events/lak/lak21/.

The book is addressed to readers that want to build a good knowledge in data analytics for e-learning. As a prerequisite, the reader should be familiar with fundamental ML and statistics concepts and experienced ML models development and software development. Considering that the final chapters discuss engagement and usability, minimal background in human-computer interaction may be helpful but not necessarily required.

The development of data analytics in the e-learning research area was accelerated by the increasing number of publicly and high-quality available data sources. From this perspective, the datasets usually have a complex structure since they represent the performed activities of various parties (*i.e.*, learners, professors, administrators, etc.) in a complex environment in which we may find disciplines, chapters, concepts, quizzes, messages, and possibly many more. This aspect increases the complexity of the data analytics problems. Still, it has the advantage of opening the way towards many issues that may be addressed and, therefore, to many approaches that may be tried. In short, the solution for solving such complex problems requires designing complex data analysis workflows which have a starting point the dataset produced by the e-learning platform along with real-time input (*i.e.*, learner's ID, learner's query, etc.) and have as output something (*i.e.*, list of quizzes, recommended tutors, etc.) that is visible in the platform and had the goal to increase the engagement and usability and finally to improve the learning experience and performance of the student.

Presented examples and approaches have several limitations. Many of the models use decision trees as the main ML algorithm, which was an informed choice due to its interpretability. There are many other fancier options (*i.e.*, Random Forest, XGBoost, Neural Networks of various types, etc.) for ML algorithms. It is up to the reader to integrate and benchmark different algorithms and solve particular problems. Another aspect not covered in this book and has gotten a lot of attention lately is computational linguistics. As deep learning algorithms are lately heavily used for solving NLP tasks through transformers, the e-learning application domain is also starting to use such techniques for tasks as text classification, group communication analysis, short answer grading, and many more.

Therefore, the distinguished chapter authors present a backbone workflow of a data analytics in e-learning, remaining to the reader—researcher or practitioner—to use, integrate, or customize described ideas for tackling their challenges faced in their e-learning applications.

Acknowledgment. This work was partially supported by project 149C/2018 "Dezvoltarea platformei e-Learning Tesys (en., Development of the Tesys e-Learning platform)" cofinanced by the University of Craiova.

Craiova, Romania   Marian Cristian Mihăescu

Contents

Introduction to Data Analytics in e-Learning

C. M. Ionașcu[ORCID], P. S. Popescu, M. L. Mocanu, and M. C. Mihăescu

Abstract The learning process is very complex. Nowadays, with the rapid development of the learning management system (LMS) platform and their increasing accessibility, any learning process has become a massive data generator. It's evaluation in terms of efficiency or performance it's still a challenge. The question "What it's needed to do to make it better adapted to the user?" has the right answer hidden into the generated data. In this context, data analytics became very useful in helping us to find the right solution and the steps to implement it in practice. In this chapter, we try to clarify what data analytics is, how it's changed to face the particularities of the e-learning process, the particularities of the e-learning process and how it can help in this specific context.

Keywords Data analytics · e-learning · e-learning analytics · Descriptive analytics · Diagnostic analytics · Predictive analytics · Prescriptive analytics

1 What is Data Analytics?

According to Popescu [33] "There is a significant amount of data around us: public or private, in numeric, text or image formats. This data needs to be considered, analysed and reviewed at a scale that exceeds the possibilities of a single person or a computing machineability. This data requires collaboration between these entities. Automatic instruments like visualization and statistics can augment the data analysis process and offer the necessary support for learning and interaction.

The collaborative processes performed around data get a different view today, and that differs from the perspective they got not long ago. These processes imply different entities: humans, computing machines, computing programs, interfaces and

C. M. Ionașcu (✉)
Department of Statistics and Economic Informatics, University of Craiova, Str. A.I. Cuza, Nr. 13, 700505 Craiova, Romania
e-mail: costel.ionascu@edu.ucv.ro

P. S. Popescu · M. L. Mocanu · M. C. Mihăescu
Department of Computer Science and Information Technology, University of Craiova, Str. A.I. Cuza, Nr. 13, 700505 Craiova, Romania

© The Author(s), under exclusive license to Springer Nature Switzerland AG 2022
M. Mihaescu and M. Mihaescu (eds.), *Data Analytics in e-Learning: Approaches and Applications*, Intelligent Systems Reference Library 220,
https://doi.org/10.1007/978-3-030-96644-7_1

projections of a system upon another. These are processes that indicate, in a significant manner, a closer view upon several aspects that were less discussed until now.

By focusing on unique aspects, i.e., on data analysis performed by trained experts, a significant potential is lost (or, to be more precisely the collaborative potential which implies simple participants is omitted), and also a significant result is lost, which is the one that refers to the collaboration that impacts the analysis results. The effort to "hire" small participants to the processes that imply these significant data quantities requires specific model design, efficient data sharing, careful interface design, improving the interaction between users, etc.

All these may seem difficult but will bring a reward that consists of efficient use of a system that takes them into account."

To understand what happened in a specific field of activity first step was always to collect data and create records. The next step was to find methods and techniques to process that data, extract useful information and then interpret that information and understand what happened. The next logical step is to build models that can be used to predict what will happen next.

Data analytics can be used to do all of that. Expressed in simple terms, the core of data analytics is the analysis of data to find patterns or trends, answer questions and draw data-driven conclusions. It is based on variety of methods and techniques from statistics, data mining and machine learning.

According to [10, 26] data analytics is a process to discover helpful information to fundament the decision-making process that have as a main goal to improve performance and efficiency of any organisation.

Because it can be easily generalized, the data analytics process can be applied to a large variety of domains.

The data analytics process is based on a series of steps like data extraction, data management, statistical analysis, and data presentation (Fig. 1).

Depending on the field where is applied and the goal of the study, the steps shown above will not have the same significance and are not equally balanced.

Data extraction is an essential step for many data analytics processes. It has as its primary purpose the extraction of the necessary data from any data sources like complex databases, raw sensor data, web text, images etc. During this step, the extracted raw data are pre-processed into useful format type, cleaned and prepared for analysis. This step usually consumes the most significant amount of time from the total time of the analysis.

In the data management step, the databases that allow easy access to the results of data extraction are prepared. SQL databases are most used type due to the fact that data management system for this database type are usually free and easy to

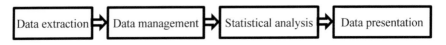

Fig. 1 Data analysis process steps

use, but nowadays the use of non-relational and NoSQL databases like document-based NoSQL databases (e.g. MongoDB, Orient DB, BaseX), key-value databases (e.g. DynamoDB, Redis, Aerospike), wide column-based databases (e.g. Cassandra, Hbase) or graph-based databases (e.g. Neo4j, Amazon Neptune) became also extent.

In the statistical analysis step, the data extracted are actually processed and used to build new information and knowledge.

The process of data is done using statistics methodology and machine learning techniques. Statistics is mostly used for transversal data analysis (for describing data, to determine variation, concentration, correlation and causality, for identifying the principal components or data dimensionality reduction) or for longitudinal data analysis when the goal is to determine the data changes in time, to identify trends of data and to create prediction models. Machine learning is mostly used for finding patterns in data, for clustering, classification, to build decisions tree etc. In this step, due to the massive amount of data, statistical programming languages such as R, Python, Octave, SAS, Stata or other dedicated programming languages with specialised packages or open-source libraries are essential [30].

Data presentation is the final step in most data analytics processes and allows to share the information obtained with stakeholders. The most often used tool in this step is Data visualization because anybody knows "an image it's easier to understand than a thousand words" and it can help anybody to understand the importance of these insights. Tools like Tableau, Looker, Google Charts, Sisense, IBM Cognos Analytics, Microsoft Power BI, SAP Analytics Cloud are used to fulfil this goal.

Usually, the data analytics process has two parts: descriptive analytics and advanced analytics. In the first part, the recorded data are described. This part has the purpose to answer the following question "what happened?". It can do that by aggregating or summarizing data in a helpful way for its user, but without making predictions. Specialized set of indicators can be calculated, and that usually is depending by the field where the analysis is done.

The second part of the process advanced tools are used to extract patterns or trends from data and to make predictions. These tools include a lot of statistics methodologies like statistical sampling, correlation, regression, analyse of variance, principal component analysis and machine learning technologies like neural networks, natural languages processing, sentiment analysis, user behaviour analysis etc. In this second part, new information from data is extracted and used to answer questions like: "what will happen?" or "what if?" (Fig. 2).

All of these are possible today because of massively growing available data sets and cheap computing power. It's also very clear that data analytics uses are almost limitless because today data are collected almost in every activity field and that creates new possibilities to apply it.

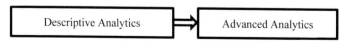

Fig. 2 Data analysis process

For example, many companies try to keep existing customers by analysing their behaviour and finding new ways to trigger specific actions. Using information gathered from the existing customers the companies also try to attract new customers.

In e-commerce the companies found out that a good idea (classical already) for maximising sales is to implement an active platform that reminds customers to finalise their shopping cart. According to Baymard Institute [34], 68% of online shopping carts are abandoned in 2021. For every 100 potential customers, 68 of them will leave without purchasing. This situation leads to significant losses.

In the software industry, companies try to keep the existing customers by implementing new ways to control interaction between them or between them and applications (apps) at high levels. In-app or clouds messages platform, apps integrated socialising platform or new ways to set the rewards appear to stimulate those interactions. Now data analytics can help to determine their efficiency too.

In manufacturing industries, companies usually record the runtime, downtime, and work queue for machines from the manufacturing line or timing for the supply chain. Using data analytics now, they can find new ways to adjust the workloads to avoid bottlenecks and to operate machines closer to their maximum capacity or to minimise the delays from the supply chain when the speed or structure of the manufacturing line is changed [14].

1.1 Types of Data Analytics

Based on the analysis of the goals of data analytics, even if it is a large subject and has a wide application field, there are identified four main types of analytics: descriptive, diagnostic, predictive and prescriptive.

Each of these types has a different goal and a different place in the data analysis process.

As already pointed, descriptive analytics helps answer questions about what happened. The techniques used in this type of data analytics summarise large datasets to describe outcomes to its users. Statistic indicators of central tendency (like average, median, mode, quantiles) and variation (like range, variance, standard deviation, variation coefficient) are often used. Specialised metrics are also developed to track performance in specific domains. Descriptive analytics requires the collection of relevant data, help to process data and provides essential insight into past performance.

Diagnostic analytics is focused on answering questions—like "why this happened?". Diagnostic analytics takes the results of descriptive analytics and continues to process them more to find what causes led to the current situation. In other words, the resulted indicators from descriptive analytics are further analysed to discover why they got higher or lower values.

Diagnostic analytics is based mainly on three steps:

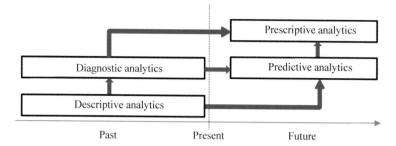

Fig. 3 Types of data analysis

1. Identify anomalies in the collected data.
2. Collect more data linked to these anomalies.
3. Use the statistical methodology to find an explanation of these anomalies by determining existing causality.

Predictive analytics is focused on finding the answer to this central question "what will happen in the future?". It uses descriptive analytics findings in the historical data to identify trends and cycles patterns and then determine what it's possible to manifest in the future. Tools used in predictive analytics are also borrowed from Statistics and machine learning.

Prescriptive analytics' main goal is to answer the question, "if this is the situation now, what should be done?".

Prescriptive analytics takes diagnostic and predictive analytics results and builds a new course of actions by proposing data-driven decisions. This type of analytics allows users in any process to make informed decisions, even in the case of uncertainty.

As in the aforementioned types of data analytics, prescriptive analytics also uses statistics and machine learning techniques. Using prescriptive analytics, the likelihoods of different outcomes are estimated by analysing past events and decisions (Fig. 3).

2 Data Analytics and Learning

The principles of data analytics can be applied across any field, including education and narrower to e-learning. It can be used to build intelligent techniques to assist students and teachers during the educational process or academic institution by analyzing the flow of knowledge and thus increasing the efficiency of the entire process. Applying data analytics to education led to the emergence of many branches like educational data mining (EDM) [6], academic analytics (AA) [12, 18], and learning analytics (LA) [12, 29].

EDM focuses on the use of data mining techniques on educational data [35]. Analysing the differences between EDM and LA, until today, a clear delimitation was not found. A different situation appears when we compare LA and AA where a clear difference between the goals of both branches emerges. LA focuses on the course level, AA focus is the institutional level. Even though all of these branches does not have the same focus, the processes used in the analysis are very similar. All of them collect and analyse data produced in the learning process and data about the learner profile and feedback to determine the learning path and use that to predict the learner's outcomes [15].

In the last ten years, the learning and teaching processes have undergone important changes generated by the new facilities offered by the technological discoveries that have become usable in all stages of these processes [16, 17].

Nowadays, LA has become one of the fields where more and more researchers and practitioners try to make new contributions [21]. There is no generally accepted definition of learning analytics, but in 2011, at the 1st International Conference on Learning Analytics and Knowledge, was defined the most widely referenced one: "Learning analytics is the measurement, collection, analysis and reporting of data about learners and their contexts, for purposes of understanding and optimising learning and the environments in which it occurs" [29]. Usually, the educational process generates a large quantity of data which contains the students' and teachers' traces in the LMS and LA it's used analyse them [2, 4]. The large availability of LMS platform helps to record this data.

LA use data analytics techniques on data collected from LMS databases. This data can be data that describe the users' interaction with the online platform and learning content (like login, logout, open a specific course, topic or content, self-evaluation using a predefined quiz, communicating with teachers or other users, accessing a forum topic, data and time for starting/stopping any specific action etc.), data that describe the users' results for the evaluation tests (like grade for each test during the course, final grade for each exam etc.) and data describing user profile (such gender, age, location, study program, year of study, last educational level before the current course, preview experience and knowledge regarding the currently studied topic etc.).

LA is an iterative data-driven process with the following goal: model, evaluate and improve the efficiency of a learning process. These data-dependent models track, measure, analyse and report findings using high numbers of the learning content dimensions (Fig. 4).

More than that, LA can be extended by using data about students' behaviour (with their agreement) from other additional sources not included in an LMS platform like social or professional networks, blogs, forums, etc.

LA can contribute to:

– *Enhancing the understanding of learning behaviours* [36].

An example of better understanding of learner behaviour in e-Learning is based on learners' online access to the LMS platform and how they engage in activities. The way in which learners engage with the activities in the LMS is unique and creates a

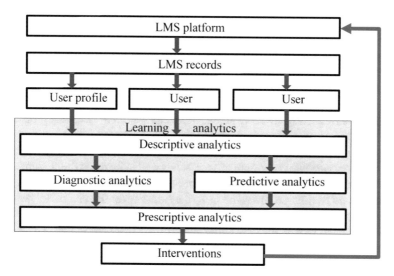

Fig. 4 Learning analytics used with an LMS platform

digital footprint that can be identified and evaluated. In an LMS, materials available online to learners are usually presented in different forms such as recorded video or audio lectures, text, quizzes, presentations, etc. and accessed by learners in a sequence that is always found between the beginning and the end of a course. An access to an online material by a learner triggers a learning process, and this is logged by the LMS in the form of an interaction carried out at a specific point in time. Every interaction made by a learner (such as: a learner accesses a certain content or watch a video) is recorded as an instance, i.e., usually a row in a table contained in a dedicated database. All of these interactions can be characterized by measurable attributes that describe how recently (immediacy of access), how frequently (frequency of access), and how long the interaction took place (duration on task). These attributes can be used to understand more clearly the enthusiasm, the determination with which the learner learns. To determine the values of the above-mentioned attributes, the two moments that delimit the learning activity, the date that marks the beginning and the end of a course, are used as a reference.

Today's learning activities are learner-centred. Activities created by teachers and uploaded into an LMS are built on the idea that the learner will play an active role in the learning process, actively use the materials, interact with teachers and other learners to understand more easily, assimilate complex knowledge, understand the meaning of the assimilated information and use it for problem solving [24]. In an LMS all the activity performed by each learner are logged and can be analysed to determine their level of active involvement in the learning process and this is usually strongly correlated with the value of the immediacy attribute. In other words, a highly engaged learner will also have a low reaction time when new resources and activities become available in the LMS and will access them faster than a less active learner.

The second attribute, frequency of access to learning content in an LMS, can be described as a measure of regularity of access and is also a key variable strongly correlated with learners' learning performance. High values of this attribute are correlated with high values of course attendance rate and ultimately with grades, according with [32]. This attribute can be also used to predict learner engagement and ultimately the learning performance. Learners who access learning materials in an LMS more frequently are more active, more engaged in the learning process, and usually assimilate knowledge faster and achieve better results. This attribute could be measured as the number of accesses during a course completed by each learner.

Another key attribute is duration of access. It is well known that the amount of time invested in learning something is crucial element of the learning process [1]. When learners use more time on the same task, they can repeat the activities, they have time to evaluate the results, learn what they did right or wrong, learn from mistakes and allow them to look at the new knowledge from a new perspective. The learning contents are usually built on the idea of keeping the learners in touch with the new knowledge sufficient time and allowing them to assimilate because the assimilation of new knowledge need time to be done.

The duration of access is also highly correlated with learner engagement. In an online environment, as in a case of LMS, this attribute can be measured by summing the of login time at the level of each course and user [19]. It needs to avoid the time recorded when the learner login and do nothing in LMS.

– *Providing useful suggestions for policymakers, teachers, and learners* [22]

LA can be used by educational institutions to improve their decision-making process in order to have greater accuracy in predicting learners' academic performance and to achieve a more efficient allocation of resources. It also can be used to help teachers to become more effective in developing teaching strategies and better content for learning activities [28].

The analysis of log data during the course from the LMS allows to identify at-risk learners in order to provide real-time support and guidance [8]. Early prediction of learning outcomes is already used in high-performance LMS platforms such as Blackboard. Early warning systems based on information extracted from LMS logs can be implemented to help detect hard-to-understand concepts or difficult activities for learners in different at-risk categories, to alert teachers and educational institutions and possibly intervene during the learning process to provide personalized learning support and to improve learning performance. For example, using LA results, clusters of different types of learners established by various and relevant criteria can be built and this will allow teachers to adopt different strategies for different type of learners.

– *Helping educational practitioners to improve teaching and learning effectiveness* [9] *and together with developers to build better learning environment such as LMS.*

Recent technological developments and advances in the understanding of the learning process allow today's education to become increasingly personalized. The role of the teacher is changing from that of a creator of learning content to that of a designer of learning scenarios that are increasingly tailored to a group or individual learner. Although still the educational institution is that sets the parameters of the course, the teacher is the one who sets the learning objectives for the learners in the course, identifies the relevant resources and creates the necessary experiences by assembling the educational content to achieve those learning objectives as effectively as possible. Instead of providing exactly the same content for all learners and the same topics, the teacher can provide a wider set of resources for learners to explore at their own pace, each building their own learning map on the fly.

This new learning philosophy is being taken up and translated by developers into new LMS functionalities.

LA allows, through the analysis of the data logged by LMSs on learners' activity, the extraction of all the information needed, on the one hand by teachers to create learning resources and on the other hand by developers to improve LMS.

If we dig deeper, we find that LA is used for: understanding the current status of learning, measuring teaching and learning efficiency, comparing the use and efficiency of different learning technologies, visualising and interpreting data, finding the remedial interventions to the learning environment [3, 5, 37].

In summary, LA systems can be used to develop models and apply them to answer to the following questions: When learners are ready to move on to the next learning concept or topic? When learners are falling behind in a course? When a learner is at risk for not completing a specific course? What grade a learner is likely to get without or with intervention? What can be the best intervention for a specific learner in order to help him during the learning process? What is the best next course for a specific learner?

Used often in e-learning, Data Analytics became specialised and received a new name: e-Learning Analytics (eLA).

Usual the final goal of eLA is to improve the students' and tutors' performance. More precisely, it is used to:

1. Reduce students' dropouts [31]. There are many courses with a higher level of difficulty that can determine students to abandon learning. Using e-learning analytics, we can determine precisely in which part of the course, at what concept students lose their focus and stop learning. Knowing this information teachers can avoid or minimise this by building the training material using new innovative ways.
2. Improve students' understanding and learning. Using eLA we can determine the learning path for each user, determine clusters of similar users, and build a concept map of each training material to find a better-adapted way to each user of user group learning path.
3. Decide which content is relevant for a given student. In the learning process exploring additional information in order to understand a new concept often occurs. eLA can help in this case through the analysis of the user historical behaviour, previous used learning material, determining connected content, and identifying if that content is more relevant for the given student.

4. Improve training materials. There is no perfect recipe for building a training material. Often this process is a long iterative process. eLA can speed up this process by assisting teachers by providing them helpful information in different adapted ways to improve the learning materials. This can be done at the level of different student types.

5. Enhance tutoring capabilities. Working in online environment with a large number of students it's not easy. This situation often occurs in practice because of efficiency reasons. Teachers and students need assistance in the learning process. On one hand, eLA can help teachers by providing information about the activity and performance of the students, to address clusters of students, by highlighting the easiest path to learn a specific concept, and on the other hand, can help students to interact with the learning content and to collaborate with the other students.

6. Improve the interaction with e-learning platforms and with the other students. eLA can help not only users of LMS. It can also help the developers improve the structure, interface, and features of LMS platform to make easier the interaction of the users with the platform.

7. Reduce tutoring costs. By helping to find the fastest way to learn some specific content, eLA can speed up the learning process and allows to repeat the same process with a large number of students using the minimum number of human resources. This is the base that Coursera, edX, future Learn or Udemy stand on.

3 Limitations of Learning Analytics

LA is neither a free or perfect tool, it has a series of limitations which involve costs, technical challenges, institutional capacity, legal and ethical issues.

As previously mentioned, online learning technologies create opportunities for researchers and developers to develop personalized learning environments based on the analysis of students' online activity for continuous improvement. All the advantages created by these technologies are dependent on the real technical possibilities to record in real time the necessary data from many students simultaneously. It is a real challenge for organisations that want to use these technologies to successfully implement LA. As seen above, LA is based on techniques for extracting educational data, usually from an LMS, and analysing it, and this means that the institutions concerned must have sufficient resources, local or remote, to record, store, protect, use and analyse large volumes of data, often including personal data, and not least to bear all the costs of the software services. The technologies on which LA relies can be costly in economic or technical terms, and this problem cannot always be solved at the level of any institution wishing to implement it. An eloquent example of this is revealed by the effects of the COVID-19 pandemic which on the one hand highlighted the need for their use, and on the other hand highlighted the efforts required and the resources that need to be covered in a relatively short time.

Based on these challenges, a number of important issues need to be considered when implementing data mining and analysis using LA. These include the clear synthesis of the questions, the choice of the questions to be answered by the LA and the selection of the necessary data to be collected. Developers need to think strategically about the data collected to answer the most important questions. LA, due to the difficulties listed above, cannot always be used to answer to all desired questions.

The lack of interoperability between different data systems is a challenge for extracting and analysing the data needed for an LA, especially when relying on different distributed stored data. Because not all hardware and software acquisitions are made in the same period of time, but usually in several successive periods, fragmentation can occur which causes the impossibility of recording or limited recording of the necessary data or only some.

Researchers in the field of online education try to extract useful information about the topics or concepts learned by a student based on his/her interaction with an LMS. This information forms the basis for conclusions and subsequent decisions to intervene in and personalise the learning process. A problem is generated by the fact that these conclusions can be confirmed by comparing assessment results and course grades with or without learning environment personalization. Moving from data extracted from multiple sources and then combining them to synthesise conclusions that characterise students' behaviour in an integrated and comprehensive way is not easy. The results of teacher-created test assessments may not be directly comparable, given the diversity of ways of assessing and quantifying them. If the LA system is implemented "on the fly" it can frequently end up in a situation where existing data sets have not been designed to simplify profiling of student behaviour. There may be missing data that could be very important in creating such models.

Technical limitations can be overcome with the help of technological progress and that takes time. The costs of storage and processing power required to process and protect data can be budgeted as part of an educational institution's infrastructure costs. In addition to these costs, implementing an LA system has other costs. For example, important and specialised human resources are needed for data preparation, processing and analysis. Existing data systems at that educational institution need to be integrated, and the demands and volume of these types of activities may exceed the capacity of the existing information technology department from that institution. It is common knowledge that usual over 70% of the effort required for data analysis is consumed by data cleaning, transformation and synchronization.

If technical problems can be overcome and data could be prepared and analysed, there is still a need for an intelligent way to present and use data. Teachers and educational decision-makers have access to many reports generated through data analysis and often are overwhelmed by the effort of selecting the really useful ones. Data dashboards need to be tailored to a simple user perspective to allow the information presented to be understood. There are usually three perspectives in creating such a system: one from the developers, a second from the administrators and last but not least, a third from the simple users. LA software developers need to understand the users' perspective of the reports very well in order for the resulting LA system to be successfully implemented. For the correct understanding of the other two perspectives as well, software developers must ask those questions that matter to teachers and administrators and integrate facilities into the LA that answer these questions.

There are two views on how to collect educational data. One is held by experts in the field of learning analytics who prefer a top-down approach. That is, they start from the significant questions that the LA system needs to answer in order to determine what data needs to be collected and analysed.

The second view is that of data mining researchers who prefer a bottom-up approach, i.e., a more comprehensive data collection strategy so that later on, starting from the main requirements, data exploration can be done by several methods. In this way unexpected patterns and models can be determined, that initially would not be possible to discover through a top-down approach.

Both views can be combined into one if the first view is used to define the initial characteristics of an LA and the second view is systematically applied to a randomly drawn sample of users to collect additional data whose exploration will lead to further improvements of the LA.

Another limitation may be generated by the nature of educational data which is usually heterogeneous and presents a hierarchical structure.

In order to perform a quality analysis, it is essential to identify those data structures and formats that accurately represent each event in the learning process of the students, targeted by the analysis. The best data structures that should be used are always conditioned by the types of problems to be solved. Very often the answer to a synthetic question can be found by extracting and preparing a large volume of data and using a complex and powerful processing algorithm. Other types of questions are, for example, aimed at determining atypical student behaviours (such as: non-rhythmic learning, errors that students make at the beginning of the process of assimilation of concepts, etc.). Detecting these behaviours is based on determining outliers from a data set, does not require laborious preparation of the data for processing and can be based on a simpler algorithm.

It is a great advantage that among the beneficiaries of the information generated from the LA there are so-called "smart data consumers", users (teachers, educational institution administrators, students etc.) who understand data and also have the perspective of the beneficiary of information and reports. A good smart data consumer needs to have an open mind to what the data is saying and have a good understanding of the reality described by data. It is possible that after the extraction and analysis of educational data, conclusions drawn to confirm or not teachers' and students' beliefs regarding the knowledge, skills and effort put during the educational process by students, but not always. For example, teachers may underestimate or overestimate the learning capacity of some students, but students may also report that they devote more time and effort to learning than they actually do.

Smart data consumers are extremely useful in the process of identifying the questions that need to be addressed, the data needed to answer those questions, how to construct and present the reports generated to become more meaningful and useful. They can help to interpret data, to identify patterns, the meaning of the patterns and the category they belong to and moreover to guide developers more quickly towards building really useful models. It should not be forgotten that in the end the technology provided by an LA has a supporting role in the essential human and social effort to make sense of experience and not a replacement for the human factor.

Working with personal data can creates privacy and ethics issues.

It has been known for many years that to succeed in creating a personalized interaction based on the generation of a user model has significant privacy implications (e.g., [27]), because it is necessary to collect personal information of students in order to personalize software according to their characteristics. Over time, quite a few issues in the area of personal data protection have been highlighted. Aggregated information from multiple data sources (LA, public social interaction networks, communication networks, etc.) can lead to information that violates the right to privacy [7].

When an educational institution has implemented the LMS and the LA using its own servers, the institution it that has to ensure the implementation and compliance with the standards regarding security and protection of personal data throughout the chain of collection, storage, retrieval and analysis. If, however, it purchases an analytics-based solution that is performed or only hosted externally by a third party, then personal data must be very clearly identified and rules imposed which the third party must follow when data is analysed and reports are published. Due to the complexity of the possible analyses to be carried out and the large volume of data to be analysed, it is possible that breaches of personal data protection standards may occur. As a rule, the more characteristics of students and teachers are analysed, the more likely it is that such situations will occur.

Educational data mining and learning process analysis allow modelling, prediction and recommendation generation, all made possible by the fact that the LA and the analysts who use it have access to records of students' actions, and students' personal data, and it's necessary that both must meet ethical and social standards. Educational data analysts must disseminate the information they discover to those who can benefit from it, and what is disseminated must be generating benefit and not creating harm.

The way in which performance evaluation is carried out and the dissemination of results to beneficiaries generates an impact on students but also on teachers' activities. Before any decision is taken, decision-makers and educational policy-makers must ensure that each predictive model or report used in the decision-making process is valid. Decision-makers need to be able to explain the issues underlying the predictions and actions automatically made by the LA based on analysis of educational data. Whenever patterns or associations involving sensitive information are discovered through the analysis of educational data, it is necessary to validate them by collecting and analysing additional data.

4 Future Challenges

As has already been the case in other fields, significant progress has been made in the relatively new field of LA in recent years. However, even as it is boosted by the development of technology, LA still faces significant challenges such as:

Building a stronger connection with the learning sciences.

The process of understanding and optimising learning is still in its infancy and requires as solid a grounded understanding as possible of how it is achieved and how it can be sustained. Solutions have been proposed so far to help understand the learning process, such as detailing educational experiences, analyzing content usage, identifying common process errors using tools like Frequent Pattern Mining [20], identifying key users using social network analysis [23] etc.

Development of analysis methodology using different data sets to optimise learning environments.

The process of making learning environments more open, more informal and more adapted to the individual needs of users is gradually leading to a move away from the use of typical LMSs by students and teachers. In order to make this process possible, the use of increasingly complex and heterogeneous datasets is required, which may include mobile, biometric and even affective data.

Methods for working with diverse datasets may include among others the implementation of ETL-like tools [25] to support different types of sources, tools to transform events into a common format such as xAPI [38], client libraries (JS) to capture events in content or client platforms, and browser extensions to capture events from any web source.

Focus on the student.

The main focus of LA is on the learning process and for this reason it is necessary to focus first on the beneficiary of the learning process, the needs of the student and only then on the needs of the institutions. For this reason, it is desirable to extend the already known criteria for a successful learning process beyond grades and persistence to include new variables such as motivation, confidence, enjoyment and achievement. Algorithms have been developed that can characterize the beneficiary of the learning process based on its associated characteristics (such as: empowerment, social, performance, effort, involvement), that can propose activity recommendations, group work recommendations and possible intervention alerts when the learning process is taking place outside the designed parameters, either for the direct use of the student or as support for the teacher in defining the action strategy for risk situations such as: school failure, reduction of the intensity of the educational process.

Develop and apply a clear set of ethical standards.

Collection of student data and further processing must be carried out with the explicit consent of students and be properly justified. So far, there are no internationally agreed standard procedures for reporting collected data, consenting or limiting their use. This issue is in the process of being regulated. For this reason, the rights of students and the responsibilities of organisations using LA processes need to be clearly and concisely defined.

Closely related to the above are inevitably other technological challenges.

The new data analysis methods used in LA are driven by an increasing volume of data. Today's technological developments lead to an increase in the volume of data frequently collected, and this volume of data is growing at such a rapid rhythm that it is difficult for institutions to keep up. This situation was accelerated in the last two years because of the effects of COVID-19 pandemic. Storing, managing and analyzing this voluminous amount of data is a major challenge [39]. Most enterprises have faced problems with their data analytics projects. A similar situation occurs in educational institutions.

In summary, the technological challenges that institutions will continue to face in the future include:

Faster management of increasingly large volumes of data.

It is already a known fact that the volumes of data collected are growing at an increasingly rapid pace. Managing large and growing volumes of data is becoming a significant challenge, as the volume of data stored in the world's IT systems is expected to double every three years [13]. Much of this data is semi-structured or unstructured. Data from different sources such as sensors, social media, news feeds, web logs, documents, emails, images, audio or video files and other types of unstructured data are still difficult to systematise and analyse. It is always desired that the decision-making process, wherever it takes place (in companies or in educational institutions), to be efficient, but to make and maintain such a desire possible requires a strong infrastructure, adaptable to the rapidly growing volume of data and able to process the data quickly to provide the information needed for decision-making, if possible, even in real time. In order to cope with the increasing volume of data, to manage the problems of storing data and processing different types of data to extract the information they need, companies and institutions frequently turn to platforms specialised in analysing large volumes of data.

Data privacy and security.

Data security is another major concern for institutions and companies with large volumes of data. Locations where large amounts of data are stored have sooner or later become hot spots for hackers and need to be protected to face persistent security threats. Increasing compliance requirements with current data security and privacy regulations, together with growing public awareness of data privacy concerns, puts even more pressure on institutions and increases the importance of managing data securely. As the volume of collection, access, storage operations increase, some organisations find themselves unable to increase the volume of regular verification and maintenance operations at the same rate. In addition, current data privacy regulations differ from one geographic location to another and are constantly changing. They apply differently depending on the type of data and its intended use. It is essential for any institution to ensure and maintain data security as a priority for all data analysis solutions that are or will be used right from their design phase and to select those technology and service providers that ensure security to the same standards as the recipient institution.

Extract meaningful information from large volumes of data.

It goes without saying that institutions aim not only to store large volumes of data, but also to use it to achieve their objectives. Therefore, institutions need to extract relevant information from the collected data, and their decision-makers need to have access to good quality information when they need it in their decision-making process. This 'democratisation' of data and the provision of self-service reporting and analysis capabilities are major challenges facing institutions seeking to become more data-driven. To achieve the required speed, the trend for institutions is to turn to the latest analytical tools, which further reduce the time needed to generate information and reports. This requires the right infrastructure and data sources to enable critical analysis, i.e., creating the necessary prerequisites to be sure that the whole process is in scope.

Increasing cost for data handling.

Handling the large volume of data, from the adoption phase to the actual implementation, requires an adequate infrastructure, which can be achieved through investment. While big data offers great benefits [11] for LA, it also generates additional costs related to the development, installation, configuration and maintenance of the necessary infrastructure, additional barriers that institutions need to overcome.

Efficient planning, done correctly and taking into account current and future infrastructure needs, will always help to limit or eliminate unnecessary expenditure.

Scaling issues.

A unique feature of the data, already mentioned, is that its volume has only one trend, namely accelerating growth. This characteristic raises a further issue, namely that the analyses to be performed on the data collected and used in the LA must also scale at the same rate, so as to keep pace with the growth in data volume. This scaling up from the perspective of many institutions is not easy to achieve. Similarly, information collection and reporting are becoming increasingly complex. Creating a flexible system that can always scale over time to accommodate organisational growth is essential and increases performance within the system. A modern data system architecture can help support and manage any changes related to data growth with greater ease and speed while doing so in a cost-effective manner.

5   Conclusions

Data analytics is a complex process of data processing based on statistical and computer science methodology, which arose from the need to understand the reality of a particular field of activity, often using large volumes of recorded data about that reality. It has multiple objectives, depending on the purpose for which it is used but also on the domain in which it is used. It can be used to describe the situation existing at a given time in an activity field, to identify the causes of the state reached, to predict the next state and last but not least to discover the measures to be applied to reach a desired state.

Data analytics combines statistics and informatics in new ways to achieve its goals. It is highly adaptable and can be used in almost all activity fields. It allows the synthesis of information often impossible to obtain by other methods. Because it has proven to be very useful and has unique advantages, it has become interesting to many researchers and practitioners who have developed it by applying it to different fields of activity, and this process has led to the emergence and development of many different, specialized branches of it.

Nowadays, the use of data analytics has become mandatory, due to its advantages, due to the increase in computer performance but also due to the increase in the volume of data accessible in almost all fields of activity.

Data analytics has been and is used in the educational field. Its development has allowed the emergence of three distinct branches only in this field until now. Thus, today there is educational datamining, academic analytics and learning analytics, and the development process is still ongoing.

Data analytics is useful for all actors involved in the learning process (teachers, students, decision-makers in educational institutions, LMS platform developers, etc.) and helps each of them to better fulfil their role. It allows teachers to find new ways of customising educational content to the characteristics of students, it allows to accelerate and monitor the educational process, i.e. to warn teachers and decision makers when it deviates from the established set of parameters, it allows to discover the decision chain that helps educational institutions to better achieve their goals and it also allows educational platform developers to develop features that meet the needs of students, teachers and educational institutions.

Putting all the above-mentioned aspects together, it is clear today that there is no other way to make learning faster, easier, more efficient and more tailored to users' needs than using LA.

While all this is possible with LA, these advantages do not come without any cost. Educational institutions that want to implement LA-based systems need to take this into account, as implementing an LA requires significant technical, human, financial and time resources.

Ethical and personal data protection issues are also not to be neglected, as accessing and processing increasingly complex volumes of data makes it more and more likely that personal data will be made public unintentionally.

References

1. Albert, S., Kussmaul, C.: Why wikis are wonderful for writing. In: Carter, T., Clayton, M. (eds.) Writing and the iGeneration: composition in the Computer-Mediated Classroom, pp. 49–67. Fountainhead Press Southlake, TX (2008)
2. Aldowah et al.: Educational data mining and learning analytics for 21st century higher education: a review and synthesis. Telemat. Inform. **37**, 13–49 (2019)

3. Avella, J.T., Kebritchi, M., Nunn, S., Kanai, T.: Learning analytics methods, benefits, and challenges in higher education: a systematic literature review. Online Learn. **20**(2), 13–29 (2016)
4. Baker, R.S., Inventado, P.S.: Educational data mining and learning analytics. Larusson, J.A., White, B. (eds.) Learning Analytics, pp. 61–75. Springer, New York, NY (2014)
5. Baker, R., Siemens, G.: Learning analytics and educational data mining. In: Keith-Sawyer R. (ed.) Cambridge Handbook of the Leaning Sciences, 2nd edn, pp. 253–272. Cambridge University Press, New York, NY
6. Baker, R.S., Yacef, K.: The state of educational data mining in 2009: a review and future visions. J. Educ. Data Min. **1**(1), 3–17 (2009)
7. Balduzzi, M., Platzer, C., Holz, T., Kirda, E., Balzarotti, D., Kruegel, C.: Abusing social networks for automated user profiling. In: 13th International Symposium, RAID 2010 Proceedings of the Conference: recent Advances in Intrusion Detection, Ottawa, Ontario, Canada (2010). https://doi.org/10.1007/978-3-642-15512-3_22
8. Barneveld A., Arnold, K. E., Campbell, J. P. (2012). Analytics in Higher Education: Establishing a Common Language, ELI Paper 1: 2012 [online]. https://www.researchgate.net/publication/265582972_Analytics_in_Higher_Education_Establishing_a_Common_Language (accesed 30.09.2021)
9. Bienkowski, M., Feng, M., Means, B.: Enhancing teaching and learning through educational data mining and learning analytics: an issue brief, vol. 1, pp. 1–57. US Department of Education, Office of Educational Technology (2012)
10. Brynjolfsson, E., Hitt, L.M., Kim, H.H.: Strength in numbers: how does data-driven decision-making affect firm performance? (2011). https://ssrn.com/abstract=1819486. https://doi.org/10.2139/ssrn.1819486
11. Chen, C., Zhang, C.Y.: Data-intensive applications, challenges, techniques and technologies: a survey on big data. Inf. Sci. **275**, 314–347 (2014)
12. Chatti, M., Dyckhoff, U., Schroeder, U., Thüs, H.: A reference model for learning analytics. Int. J. Technol. Enhanc. Learn. (IJTEL). Special Issue on State-of-the-Art in TEL, 318–331
13. Coyle, D., Li, W.: The data economy: market size and global trade. ESCoE Discussion Paper No. 2021-09 (2021). https://escoe-website.s3.amazonaws.com/wp-content/uploads/2021/08/02103632/ESCoE-DP-2021-09.pdf. Accessed 03 Nov 2021
14. Darvazeh, S.S., Vanani, I.R., Musolu, F.M.: Big data analytics and its applications in supply chain management. In: Martínez, L.R., Osornio Rios, R.A., Prieto, M.D. (eds.) New Trends in the Use of Artificial Intelligence for the Industry 4.0. IntechOpen (2020). https://doi.org/10.5772/intechopen.89426. https://www.intechopen.com/chapters/69320. Accessed 22 Aug. 2021
15. Diaz, V., Brown, M.: Blended learning: a report on ELI focus session. EDUCAUSE Learning Initiative (ELI), ELI Papers and Reports (2010). https://library.educause.edu/~/media/files/library/2010/11/eli3023-pdf.pdf. Accessed 27 Aug. 2021
16. Ferguson, R.: Learning analytics: drivers, developments, and challenges. Int. J. Technol. Enhanc. Learn. 304–317 (2012)
17. Fu, Q.K., Lin, C.J., Hwang, G.J.: Research trends and applications of technology-supported peer assessment: a review of selected journal publications from 2007 to 2016. J. Comput. Educ. **6**(2), 191–213 (2019)
18. Goldstein, P.J., Katz, R.N.: Academic analytics: the uses of management information and technology in higher education. EDUCAUSE Center for Applied Research (2005) https://er.educause.edu/-/media/files/articles/2007/7/ekf0508.pdf?la=en&hash=72921740F4D3C3E7F45B5989EBF86FD19F3EA2D7. Accessed 20 Aug. 2021
19. Gunn, A.M., Richburg, R.W., Smilkstein, R.: Igniting Student Potential: teaching with the Brain's Natural Learning Process. Corwin Press, Thousand Oaks
20. Han, J., Cheng, H., Xin, D., Yan, X.: Frequent pattern mining: current status and future directions. Data Min. Knowl. Disc. **15**, 55–86 (2007). https://doi.org/10.1007/s10618-006-0059-1

21. Hui, Y.K., Kwok, L.F.: A review on learning analytics. Int. J. Innov. Learn. **25**(2), 197–222 (2019)
22. Hwang, G.J., Hung, P.H., Chen, N.S., Liu, G.Z.: Mindtool-assisted in-feld learning (MAIL): an advanced ubiquitous learning project in Taiwan. Educ. Technol. Soc. **17**(2), 4–16 (2014)
23. Jan, S.K., Vlachopoulos, P., Parsell, M.: Social network analysis and learning communities in higher education online learning: a systematic literature review. Online Learn. J. (2019). https://doi.org/10.24059/olj.v23i1.1398
24. Keengwe, J., Onchwari, G., Agamba, J.: Promoting effective e-learning practices through the constructivist pedagogy. Educ. Inf. Technol. **19**, 887–898 (2014). https://doi.org/10.1007/s10639-013-9260-1
25. Kherdekar, V.A., Metkewar, P.S.: A technical comprehensive survey of ETL tools. Int. J. Appl. Eng. Res. **11**(4) (2016). https://doi.org/10.37622/IJAER/11.4.2016.2557-2559
26. Kiron, D., Shockley, R., Kruschwitz, N., Finch, G., Haydock, M.: Analytics: the widening divide. MIT Sloan Manag. Rev. (2011). http://www.greatlakessoft.com/divide/IBM_analytics_the_widening_divide_original.pdf. Accessed 22 Aug. 2021
27. Kobsa, A.: Modeling the user's conceptual knowledge in BGP-MS, a user modeling shell system. Comput. Intell. **6**(4), 193–208 (1990). https://doi.org/10.1111/j.1467-8640.1990.tb00295.x
28. Lee, L.K., Cheung, S.K.S., Kwok, L.F.: Learning analytics: current trends and innovative practices. J. Comput. Educ. **7**, 1–6 (2020). https://doi.org/10.1007/s40692-020-00155-8
29. Long, P., Siemens, G.: Penetrating the fog: analytics in learning and education. Educ. Rev. **46**(5), 31–40 (2011)
30. Najafabadi, M.M., Villanustre, F., Khoshgoftaar, T.M., Seliya, N., Wald, R., Muharemagic, E.: Deep learning applications and challenges in big data analytics. J. Big Data **2**(1) (2015)
31. Na, K.S., Tasir, Z.:. A systematic review of learning analytics intervention contributing to student success in online learning. In: 2017 International Conference on Learning and Teaching in Computing and Engineering (LaTICE), pp. 62–68 (2017). https://doi.org/10.1109/LaTiCE.2017.18
32. Piccoli, G., Ahmad, R., Ives B.: Web-based virtual learning environments: a research framework and a preliminary assessment of effectiveness in basic IT skills training. MIS Q **25**(4), 401–426 (2001)
33. Popescu St.: Metode si instrumente de analiza a datelor pentru îmbunătățirea design-ului interacțiunilor in sistemele de e-learning. Ph.D. thesis (2019)
34. Reasons for cart abandonment—why 68% of users abandon their cart (2021 data). Baymard Institute (2021). https://baymard.com/blog/ecommerce-checkout-usability-report-and-benchmark. Accessed 30 Sept. 2021
35. Romero, C., Ventura, S.: Educational data mining: a survey from 1995 to 2005. Expert Syst. Appl. **33**(1), 135–146 (2007). https://doi.org/10.1016/j.eswa.2006.04.005
36. Wong, A., Chong, S.: Modelling adult learners' online engagement behaviour: proxy measures and its application. J. Comput. Educ. **5**(4), 463–479 (2018)
37. Wong, B.T.M.: Learning analytics in higher education: an analysis of case studies. Asian Assoc. Open Univ. J. **12**(1), 21–403 (2017)
38. Xiao, J., Wang, L., Zhao, J., Fu, A.: Research on adaptive learning prediction based on xAPI. Int. J. Inf. Educ. Technol. **10**(9), 679–684 (2020)
39. Zhi-Hua, Z, Chawla, N. V., Yaochu, J., Williams, G. J.: Big data opportunities and challenges: discussions from data analytics perspectives. IEEE Comput. Intell. Mag. (2014). https://doi.org/10.1109/MCI.2014.2350953

Public Datasets and Data Sources for Educational Data Mining

P. S. Popescu⬛, M. C. Mihăescu, and M. L. Mocanu

Abstract Datasets are the starting point for any Machine Learning or Data mining workflow, and their impact on the overall performance of the whole system is vast. The data sources offer a variety of data that can be used directly or needs more or less preprocessing to produce a suitable dataset, but the main problem is what data is available and what data should be used to solve a specific task. The datasets explored in this chapter focus on the educational area and can directly impact any educational environment: online or just classical full-time education. This chapter aims to clarify what public datasets we have at this point, for what tasks are suitable, and presents a use case for a specific dataset in detail. The dataset referred deeply in this chapter was recently produced and includes attributes that can easily be mined from any e-Learning platform, so it can be a good baseline for anyone who starts in Educational Data Mining or tries to perform sample experiments in order to get a better insight before implementing a workflow in their system.

Keywords Machine learning · Educational data mining · Decision trees

1 Introduction

Datasets and data sources are one of the most critical aspects of the Educational Data Mining research area, being indispensable for machine learning models and are essential factors in building successful, intelligent systems. In most systems that rely on machine learning and data mining algorithms, datasets and data sources are crucial, providing the basis for training the algorithm, but datasets involved in Educational Data Mining have unique particularities. Usually, when dealing with datasets from the education, we have a low number of instances that share similar attributes, many missing values and, in many cases, not relevant attributes for the task. Various tasks can be solved using the proper dataset combined with a carefully chosen algorithm, including prediction of final grade or failure, computing trends, grouping

P. S. Popescu (✉) · M. C. Mihăescu · M. L. Mocanu
Department of Computer Science and Information Technology, University of Craiova, Str. A.I. Cuza, Nr. 13, Craiova, Romania
e-mail: stefan.popescu@edu.ucv.ro

© The Author(s), under exclusive license to Springer Nature Switzerland AG 2022
M. Mihaescu and M. Mihaescu (eds.), *Data Analytics in e-Learning: Approaches and Applications*, Intelligent Systems Reference Library 220,
https://doi.org/10.1007/978-3-030-96644-7_2

learners based on common characteristics, or providing recommendations. Each of these tasks is different, and they need a specific dataset and a specific algorithm that can fit the data and provide a good enough accuracy; if an interpretable model can also be built, there is a plus. The interpretable models are essential in Educational Data Mining as it is presented in Chap. 4, enabling teachers and data analysts to better understand students in two ways: as a whole when referring to a group of students and individuals. Both these approaches can reveal necessary information; the first one (as a whole) refers to analyzing the model or the main characteristics of the most important instances from the model in order to understand the students better and to use the information as feedback and the second one refers individual analysis which means that professors need to analyze the place of the instance in the model and to obtain relevant information about a student in comparison to the rest. An individual analysis is relevant in the Educational Data Mining context as it can reveal a student's strengths and weaknesses and offer the premises for making recommendations to improve learning performance. However, all of this analysis depends on the datasets, how they are structured and used. Depending on the tasks, it is important also to choose the right attributes as most of the datasets available today includes more than one type of attribute, adding demographic and personal information along with academic ones. These attributes have different types of impact on the model performance depending on the task; in some cases, they improve the model's accuracy and, in some cases, may lower it. Starting from this premises, we may say that we can eliminate demographical and all the attributes that are not so relevant for the model focusing on providing better performance, but it is not every time case because these attributes can improve the model's variety making it more generic and less prone to overfitting.

A variety of attributes included in the datasets means we can tackle a variety of tasks, from predicting the final result or mainly academic performance to automated essay scoring or detecting multiple accounts. There is also a variety of data sources and providers of the datasets that offer the data in multiple formats, some of the datasets being ready to use and some requiring a lot of preprocessing. The data's sources influence the data format and the amount of work necessary to prepare the dataset, and the way the data was logged, making the dataset suitable for timeline analysis of a single instance or classification tasks following other instances. This variety of datasets and data sources which provides will be further presented in this chapter along with their benefits.

This chapter tackles two main directions based on two previously published papers. The first direction refers to the DSA student's dataset [1], a dataset designed and contributed from our university, and the second direction refers to presenting and analyzing the most important and relevant available datasets [2]. Regarding the DSA dataset, the main motivation of computing and publishing such a dataset is that we can foresee the student's failure before the exam occurs, even if this is not a revolutionary approach. The motivation is that in our situation, the student took tests before the semester end up, and they had to take over the exam after that, so the grade of the test may be a relevant indicator for their result. There is also a correlation between the exams they took and the final examination, as the questions utilized in the tests referred to the same subjects they must learn for the final exam. Yet though the subjects were the same, there are several variations between the tests they took and the final test as the tests have questions from only a part of the subjects necessary at the exam.

2 Related Work: EDM Review Papers

Among the first EDM review papers, there are [3, 4]. With 1757 citations, the first paper presents EDM as a "young research area" mainly driven by the new web-based education systems that provide the trace of learners as log files ready to be processed. The available systems were classified as commercial, free, intelligent, or adaptive, while the data mining techniques consisted of clustering, classification, outlier detection, association rule mining, sequential pattern mining, and text mining.

2.1 General EDM Review Papers

We further present the best general-purpose EDM review papers since 2010. In [5], the authors describe EDM stakeholders and objectives and types of data that may be mined. A more detailed list of EDM tasks is presented as providing feedback, offering recommendations for students, predicting student's performance, student modelling, detecting undesirable student behaviours, grouping students, social network analysis, developing concept maps, constructing courseware, planning and scheduling. At this moment, PSLC DataShop [6] was the only publicly available dataset.

Later [7], points out the new LAK conference started in 2011 and the appearance of "Big Data in Education" MOOC course by Ryan Baker offered on Coursera in 2013 and on EdX in 2015 and as MOOT.[1] Still, in terms of modelling methods, they are the same as described previously in [5].

The paper [8] focuses on clustering algorithms and understanding learning styles. A comprehensive list of clustering algorithms, along with objective and dataset descriptions for 35 papers, is provided. The data sources used for the literature search were IEEEXplore, ACM Digital, JEDM (Journal of Educational Data Mining), ProQuest Education Journals and ScienceDirect. Still, there is no information provided regarding the availability of the datasets. The search criteria are presented by describing inclusion and exclusion criteria along with primary sources.

When [9, 10] papers have been published, the EDM domain is already well-established with top-notch conferences and journals. Obtaining bibliographical

---

[1] B.D.E, https://www.upenn.edu/learninganalytics/ryanbaker/bigdataeducation.html.

references was performed by searching repositories such as IEEEXplore,[2] Elsevier Science Direct,[3] Research Gate,[4] etc. A limitation of this approach regards the need for various subscriptions (i.e., personal, institutional) to obtain the papers. This limitation has been addressed by using Google Scholar and Google Dataset Search as publicly available tools to broadly search scholarly literature. The key pointed aspects regard the evaluation of the e-Learning environment with a focus on comparative aspects from traditional classroom, improvement, and evaluation of the interaction between actors, monitoring and evaluation of teaching, evaluation of pedagogical actions, evaluation of administrative management, and evaluation of the use of multimedia resources.

Of particular importance in the area of general Educational Data Mining, review papers are the works of [11–13].

The paper from 2013 clearly states that EDM is a well-established research area with five excellent conferences and 12 top related journals. The top 10 most cited papers about EDM are used to point out the goals, stakeholders, topics of interest and classical methods. Finally, a list of 22 EDM tools is presented.

The main event spotted in the 2017 paper is the appearance and well establishment of MOOCs: Coursera, edX, or Udacity. Although MOOC challenges remain quite the same as before, they suffer some modifications from the new context. Thus, analyzing students' interactions becomes analyzing usage and engagement, FORUM data or in-video interaction.

The paper from 2020 shows an increase to 9 conferences, 18 books—from which 14 are from the last five years—a list of 11 top related journals and a list of most cited papers. It also presents a link and a short description for a list of 9 tools and 13 publicly available datasets. Finally, a list of the 16 most popular methods and 23 current applications is presented. Although this paper presents, in short, a list of 13 publicly available datasets, this review aims in getting into more details about 26 datasets in an attempt to present them in a way that is more informative for researchers that would want to dive into the area of EDM quickly.

2.2 Specific EDM Review Papers

Paper [14] presents software tools that are widely used, accessible, and compelling when dealing with EDM research tasks. General-purpose data manipulation and feature engineering tools are Microsoft Excel/Google Sheets, Python and Jupyter notebooks or Structured Query Language (SQL) and algorithmic analysis tools like RapidMiner, WEKA, SPSS, etc. KNIME, Orange, KEEL, Spark MLLib, and visualization tools like Tableau or D3.js. The list of specialized EDM applications consists of Tools for Bayesian knowledge tracing (BKT), text mining, linguistic inquiry and

---

[2] IEEEXplore, https://ieeexplore.ieee.org/Xplore/home.jsp.

[3] Elsevier Science Direct, https://www.sciencedirect.com.

[4] Research Gate, https://www.researchgate.net.

word count (LIWC), WMatrix, Coh-Metrix, Word2vec, Probabilistic topic modelling tools (MALLET, LDA package, and STM), Latent semantic analysis (LSA), Natural language processing (NLP) tool kits (Stanford CoreNLP, Python).

NLTK and Apache OpenNLP, LightSIDE, ConceptNet, AlchemyAPI, TAGME, Apache Stanbol, SNA, Gephi, EgoNet and ProM. Although there are presented 40 tools, few of them are very specific to EDM workflows.

A review of data mining processes on how students solve programming problems has been addressed by [15]. The lack of publicly available datasets is pointed out as the main limitation for performing replication studies. Explicitly, the Blackbox Data Collection Project[5] logs the data that the application sends to the server is makes it available by request without making any dataset publicly available. Finally, the presented Grand Challenges are in the same line of research with most on the EDM review papers, but with more emphasis on the need for public availability of the data. In this area, we mention the latest development by [16], which present a neural-based model for building code embeddings. We state that this progress brings the area of NLP close to computer programming with several exciting applications like automatic code review or API discovery.

A review of visual analytics of educational data has been published by [17]. Educational data visualization is of critical importance for EDM processes since it represents the outcome of the workflow and should be informative and well-integrated with the e-Learning platform and consistent with educational theory. Among the most exciting approaches, we mention understanding collaboration, visualizing an instructional design, understanding relationships, promoting reflection, understanding motivation, exploring usage patterns, or exploring learning progress and paths. The critical aspect regards designing, implementing and integrating specific EDM visualization techniques and tools by particularising a wide area of available general data mining visualization techniques.

The area of text mining in education has been reviewed by [18]. The general text mining and techniques (i.e., text classification and clustering, NLP, information retrieval, text summarization) were adapted to solve educational tasks like student evaluation, student support/motivation, analytics, question/content generation, providing feedback or building recommendation systems. Since e-Learning systems are usually rich in producing data such as forums, online assignments, essays, chats, documents, blogs, and emails, there is an ample opportunity in providing the *intelligent* character for the e-Learning system that integrates such functionalities.

2.3 Findings on EDM Review Papers

The current situation in the EDM area reveals a strong establishment of research activity in terms of e-Learning systems/MOOCs, applications, tasks, tools, methods that span a relatively wide range of conferences and journals and two professional

---

[5] Data Collection Project, https://bluej.org/blackbox/.

societies (i.e., EDM and SoLAR). The wide range of EDM applications and tasks that may be developed/integrated/improved in virtually any e-Learning system represents the main ingredient that enabled the rapid growth of this research area [19]. Still, systems and progress must be evaluated, and this may be achieved only through publicly available datasets and reproducible research. In this matter, we observe that there is no systematic review on publicly available EDM datasets, as the only mention reduces to a table in [13], which provides only the link to the publicly available dataset and a minimal description.

3 Review on Other Public Educational Datasets

On the *UCI ML repository*,[6] there are various educational datasets configured and ready to be used for experiments. Although these datasets have been available for a long time and are among the most referenced datasets in the EDM community, they differ in structure, logged features, and usefulness in terms of citations, practical usages, and tackled tasks.

University Data Set[7] is the first dataset with 285 instances and 17 attributes that were donated in 1988. Unfortunately, there is no paper to describe the dataset and, therefore, not no citation or usage of the dataset, which makes it of little interest.

Teaching Assistant Evaluation Data Set[8] is another early dataset described in [20] that gathered 1340 citations, not mainly for the quality and extensive usage of the dataset but the discussion on split selection methods. The dataset itself is relatively small, with 151 instances described by five attributes.

Student Performance Dataset[9] is described in [21], has 394 citations and consists of 649 instances described by 30 attributes from which three can be used as target class. The dataset is divided into two files which can be merged because they have the same structure: one from math classes and one from Portuguese courses. Although three features can be used as the target class, two of them can be used in the classification process and positively correlate with the last. The attributes are mainly demographic and social, but some of them are school-related, and their usage is limited to predicting the student's final grade at the end of the semester or year. One idea that needs to be mentioned is that the final grade is 0–20, resulting in a significant number of class values.

This dataset has also been uploaded on Kaggle,[10] where 305 publicly available kernels perform exploratory data analysis. Unfortunately, there is no defined any task with a specific validation metric such that there is no leader board publicly available.

---

[6] UCI Machine Learning Repository, https://archive.ics.uci.edu/ml/index.php.

[7] University Data Set, https://archive.ics.uci.edu/ml/datasets/University.

[8] Data Set, https://archive.ics.uci.edu/ml/datasets/Teaching+Assistant+Evaluation.

[9] Student Performance Data Set, http://archive.ics.uci.edu/ml/datasets/Student+Performance.

[10] Student Alcohol Consumption, https://www.kaggle.com/uciml/student-alcohol-consumption.

User Knowledge Modelling Data Set[11] is described in [22], has 96 citations and consists of 403 instances defined by five attributes from which there is only one class attribute discretized into four values. In the case of this dataset, the details are highly related to learning activities with no demographic or social information about the users. The dataset is feasible mainly for classification and clustering, and it can be used to predict learners' knowledge levels. In an unsupervised context, the students are grouped in clusters representing different knowledge levels. Otherwise, the dataset has been used in the newly proposed algorithms' experimental results to test new predictive validation metrics.

Educational Process Mining Dataset (EPM)[12] is described in [23], has 42 citations and consists of 230,318 samples described by 13 attributes computed based on a group of 115 students. The dataset has been created by logging performed activities of the students while using an educational simulator. Based on this structure, the dataset has been used in classification, regression, or clustering to predict learning difficulties, analyze structured learning behaviour, self-organizing map clustering, or discover student behaviour patterns.

Open University Learning Analytics dataset[13] is another reasonably well described standalone dataset in [24]. Among the most tackled tasks within the 60 citations, we mention early identification of at-risk students, student engagement predictions, or demographics' role in online learning. The prediction tasks were addressed by classical ML algorithms (i.e., DT, J48, XGBoost, CART, SVM, or Random Forest) and were validated by classical metrics. The greatest limitation is that the dataset consists of seven *.csv* files, which have the design of a databases schema. Thus, using this dataset may require intensive preprocessing as feature values should be computed for items (i.e., student, course, assessment) that may be considered.

Student Academics Performance Data Set[14] is described in [25], has 43 citations, and consists of 300 instances described by 24 nominal, mainly demographic attributes. The main tackled tasks regard the prediction of student academic performance in a classification context.

Another important data source is Mendeley Data Repository,[15] which has been described in detail as a platform for research data management [26].

Embeddings and topic vectors for the MOOC lectures dataset[16] has been described in [27] and consists of word embeddings and document topic distribution vectors generated from 12,032 MOOCs video lecture transcripts from 200 courses collected from the Coursera learning platform. The most important shortcoming of this dataset is that it does not include the transcripts themselves. The dataset is aimed at NLP tasks and consists of ten *.csv* files, six with data about topic vectors and four with data about word embeddings.

---

[11] User Knowledge, http://archive.ics.uci.edu/ml/datasets/User+Knowledge+Modeling.

[12] Educational Process Mining (EPM), https://tinyurl.com/y27yduo3.

[13] Open University Learning Analytics dataset, https://tinyurl.com/y6ysfant.

[14] S.A.P., https://archive.ics.uci.edu/ml/datasets/Student+Academics+Performance.

[15] Mendeley Data Repository, https://data.mendeley.com.

[16] MOOC lectures dataset, https://data.mendeley.com/datasets/xknjp8pxbj/1.

Data for: Effectiveness of flip teaching on engineering students' performance in the physics lab[17] has been described in [28] and consists of a sample of 1233 students enrolled from 2013 to 2017 who completed the subjects of Physics and Electricity. The dataset contains laboratory grades and final grades both for traditional teaching methodology and flip teaching suitable for classification and regression tasks.

Dataset for Factors Affecting Teachers' Burnout[18] has been described in [29] and contains data about a sample of 876 teachers across three Indonesian provinces that completed a printed form of a questionnaire. The questionnaire, responses and factor analysis results are available, along with a final validated model and relationship testing.

Data of Academic Performance evolution for Engineering Students[19] have been described in [30] and consists of data containing academic, social, financial information for 12,411 students described by 44 attributes in a *.csv* format. The dataset may be used for prediction, classification, and evaluation models of academic and social variables.

The KEEL dataset repository has been shortly described in [31], *focusing on the KEEL Java software tool that can be used for many* different knowledge data discovery tasks. The archive dataset consists of many datasets that are also available with the standalone KEEL[20] project homepage.

Harvard Dataverse[21] is a repository for research data containing 98,873 datasets organized on 13 subjects, such as Computer and Information Science, Engineering, Arts and Humanities, etc. One advantageous key aspect is that Harvard Dataverse has implemented functionality data viewing, exploration, and other valuable features.

HarvardX Person-Course Academic Year 2013 De-Identified dataset[22] has been described in [32], a paper with 352 citations that presents detailed statistics about online courses that produced the data. The 338,223 logged instances are defined by 20 attributes and were used for Understanding the progression of users, examining access and usage patterns or predicting MOOC performance with week one behaviour.

Canvas Network Person-Course[23] is a publicly available dataset from 2016 and has 2,766 downloads with a not published paper describing the dataset. The only known usage is at the LAVA hackathon, which is presented in a later subsection. The dataset consists of 325,000 aggregate records defined by 25 attributes. The dataset has plenty of missing values which leads to a necessity for preprocessing, and there is no explicit class attribute that makes it more feasible for clustering purposes. Each record represents one individual's activity in one of 238 courses.

---

[17] Dataset, https://data.mendeley.com/datasets/68mt8gms4j/2.

[18] Dataset, https://data.mendeley.com/datasets/6jmv43nffk/2.

[19] Dataset, https://data.mendeley.com/datasets/83tcx8psxv/1.

[20] KEEL, http://www.keel.es/

[21] Harvard Dataverse, https://dataverse.harvard.edu.

[22] HarvardX Person-Course Academic Year 2013, https://doi.org/10.7910/DVN/26147.

[23] Canvas Network Person-Course, https://doi.org/10.7910/DVN/1XORAL.

Massively Open Online Course for Educators (MOOC-Ed)[24] dataset has been described in [33], a paper with 14 citations. The dataset is mainly designed to be used for Social Network Analysis, MOOCs courses dropout rate or exploring self-regulated learner profiles in MOOCs.

CAMEO Dataset: Detection and Prevention of "Multiple Account" Cheating in Massively Open Online Courses[25] have been described in [34], a paper with 69 citations that addresses the problem of cheating as a big issue in MOOCs. Unfortunately, we could not find other papers using the dataset for data analysis, although the citations tackle the same problem. The dataset consists of four restricted files that may be accessed upon request: course listings, descriptions, and actual data in *.csv* format.

Nursing Student Data[26] and *Situated Academic Writing Self-Efficacy Scale Validation*[27] are two new datasets with the description in preparation as an unpublished doctoral dissertation of Mitchell Kim from Red River College. The first dataset has 255 observations defined by 84 variables, while the second has 807 observations defined by 29.

Early Reading and Writing Assessment in Preschool Using Video Game Learning Analytics[28] is also a recent dataset authored by Amorim Americo, with no published paper describing the dataset in detail. The dataset contains data from 331 observations (i.e., students) represented by 25 variables, among which 20 regard phonological awareness, early reading and writing games, and their scores in a standardized word reading and word writing assessment.

DataShop@CMU[29] has been described in [6] and consists of a system that gathers fine-grained (i.e., very detailed), longitudinal (i.e., per semester or academic year) and extensive 40 datasets collected from ongoing courses or external courses. The data sets may be imported or exported as XML or tab-delimited text file format and are available for download along with available applications/tools that may be used to support exploratory analyses. As the DataSchop@CMU gathered data from 6 coursed in 2010, it continuously added and updated dataset versions and analysis/visualization tools. Thus, DataSchop@CMU gathers datasets about middle school math from *ASSISTments, OLI (Online Learning Initiative)* or other educational software (i.e., Andes, Cognitive Tutor, REAP, etc.). About three datasets per year were regularly added or updated, some used in EDM related data competitions or challenges. The dataset and competitions are further discussed in a distinct section.

Particular datasets are made publicly available for EDM competitions or within crowd-sourced platforms such as Kaggle. This section presents EDM competitions at top DM conferences under the supervision of prestigious professional organizations, competitions that run within a crowd-sourced platform or as hackathons. The most

---

[24] (MOOC-Ed), https://doi.org/10.7910/DVN/ZZH3UB.

[25] CAMEO Dataset, https://doi.org/10.7910/DVN/3UKVOR.

[26] Nursing Student Data, https://doi.org/10.7910/DVN/MQ8EP0.

[27] Dataset, https://doi.org/10.7910/DVN/M07HQ7.

[28] Early Reading and Writing Assessment in Preschool, https://doi.org/10.7910/DVN/V7E9XD.

[29] DataShop@CMU, https://pslcdatashop.web.cmu.edu/index.jsp?datasets=public.

major advantage of this approach is that it provides on the spot strong evidence about the quality of the proposed model/kernel and a ranking against other competitors.

KDD Cup 2010 Educational Data Mining Challenge[30] is the first EDM competition. Although it started ten years ago [35], solutions are continuously added. The organizers offer two types of datasets: for development and challenges similar to Kaggle. The development datasets consist of two *Algebra I* datasets (2005–2006 and 2006–2007) and one for *Bridge to Algebra* (2006–2007); these datasets consists of 575, 1840, and 1146 students, which are a pretty large number. For the challenge, there are two datasets: *Algebra I* and *Bridge to Algebra*. These datasets have 3310 and 6043 students, and each of these students for any of the datasets has 20 records. The evaluation of the competition was made using the RMSE metric and based on the leaderboard.

What Do You Know?[31] is a Kaggle competition launched in 2011. The task is to predict whether the student will answer a question correctly. Sixteen features describe the dataset for training and testing files. Although there are no publicly available notebooks, the public and private leaderboards show 238 teams, indicating the challenge's quality and interest. Unfortunately, there is no research paper describing the dataset or the notebooks from the leaderboards, so there is no indication regarding the approaches taken to tackle the problem.

The Automated Essay Scoring[32] and *Short Answer Scoring*[33] are two Kaggle competitions that were launched in 2012. Although 34 and 54 kernels are currently ranked according to the same objective criteria (i.e., quadratic weighted kappa error metric), there is no available public notebook with the solution. Unfortunately, the solutions may not have been described in scientific papers, so there is no possibility of studying or improving any existing solutions.

Students' Academic Performance Dataset[34] is described in [36] and is available and ready to be used on the Kaggle platform. The dataset has 460 instances described by 16 attributes and one class attribute, which has three possible values. The dataset consists of demographic and scholarly features, and it has 186 kernels that explore the data and offer results. Based on this dataset, there are 186 available kernels and have over 25,000 downloads. For this particular dataset, there is no task associated, and therefore there is no defined validation metric. Each user may define his task and start developing a public kernel as a notebook in python or R programming languages. The most significant limitation of this dataset consists of lacking a leaderboard in which kernels may be ranked according to a specific validation metric. This dataset's general task is designed to predict the interval the students will fit at the end of the semester. The 185 kernels on Kaggle and 54 citations on GS show a reasonable large usage of the dataset with the EDM community.

---

[30] KDD Cup 2010, https://pslcdatashop.web.cmu.edu/KDDCup.

[31] What Do You Know? https://www.kaggle.com/c/WhatDoYouKnow.

[32] Automated Essay Scoring, https://www.kaggle.com/c/asap-aes.

[33] Short Answer Scoring, https://www.kaggle.com/c/asap-sas.

[34] Students' Academic Performance Dataset, https://www.kaggle.com/aljarah/xAPI-Edu-Data.

NAEP 2017: ASSISTments Data Mining Competition[35] started under NAEP in 2017. The goals of the competition that made the available dataset are twofold: to do well at the task (mostly engineering-oriented) and to have the field learn from this competition (mostly science-oriented). Seventy-six attributes describe each of the 942,817 records from the dataset, and the dataset evaluation was made using both AUC and RMSE with the best result scoring 0.2618 for RMSE and 0.9551 for AUC. Regarding the publications related to this competition, eight prior studies are presented on the competition site. The successful entries have been invited to submit their results at the EDM2018 conference and to a special issue of the Journal of Educational Data Mining.

NAEP 2019: Educational Data Mining Competition[36] continues the competition first launched in 2017. The five leaders will be featured at an AIED/EDM2020 conference workshop. Still, until then, their solutions are briefly described on the competition' site. For this dataset, it is a different approach having only seven columns for each student and 438,392 instances for training and for testing three datasets that logged the data at different time stamps. The solution evaluation was made using an aggregation score which is composed of two metrics: adjusted AUC and Adjusted Kappa. The winner solution used an XGBoost Regressor with an aggregated score of 0.5657. A significant difference compared to the previous competition is that no previous publication supports it presented on the site this time.

Kaggle: Students Performance in Exams[37] is an educational dataset available on the Kaggle platform. There are a total number of 8 attributes, five of them describing the instances and the other three which can be fed to an algorithm and the other five or be predicted as class attributes. The dataset was uploaded in 2018 on the platform and is not an extensive dataset offering the data logged for 1,000 instances. The only task defined on Kaggle regards the correlations between different attributes like *Lunch and Writing Score*. As the dataset is more engineering-oriented, it is not being supported by scientific publications. The dataset has 356 kernels (most of them about data visualization, exploratory data analysis and classification), and 14 discussions opened and therefore, we conclude that it has been intensively used. Still, as it lacks evaluation methodology and, therefore, a leaderboard, we state that this is a limitation that deprives researchers of better insight into the performance obtained on a specific task.

The CSEDM 2019 Data Challenge[38] is a competition within the 2nd Educational Data Mining in Computer Science Education (CSEDM) Workshop within the 9th International Learning Analytics and Knowledge (LAK19) Conference. The challenge was building a student model that trains previous students' programming process data and predicts if new unseen students succeed at a given programming

---

[35] NAEP 2017, https://sites.google.com/view/assistmentsdatamining.

[36] NAEP 2019: https://sites.google.com/view/dataminingcompetition2019.

[37] Students Performance https://www.kaggle.com/spscientist/students-performance-in-exams.

[38] CSEDM 2019 Data Challenge, https://sites.google.com/asu.edu/csedm-ws-lak-2019.

task. The dataset[39] is from DataShop@CMU. Unfortunately, to the best of our knowledge, the results (i.e., source code of solutions, rankings, etc.) of the competition are unavailable, and further submission is not possible anymore.

2020 CSEDM Data Challenge[40] is a competition within the 4th Educational Data Mining in Computer Science Education (CSEDM) Virtual Workshop in conjunction with Educational Data Mining (EDM 2020) conference. The draft description of the data challenge (https://tinyurl.com/y3kfu426) proposes the CodeWorkout (CWO) dataset and two possible secondary datasets—PCRS and OLI Python for usage during the challenge. All datasets store programming to process data: running attempt, compilation result, and errors obtained after submitting a solution (i.e., the source code) to the appropriate grading system. Unfortunately, there is no publicly available link to the datasets, competition results, or papers regarding the proposed models.

EdNet: A Large-Scale Hierarchical Dataset in Education[41] is a new dataset described in [37], which also has an associated competition. The dataset is based on data collected over two years by Santa, a multi-platform AI training service along with more than 780 K users in Korea accessible through with Android, IOS and web.

Four smaller datasets compose the dataset: KT1, KT2, KT3, and KT4 with different extents and each of the smaller datasets are very well documented. The datasets have a different approach offering the logged actions rather than a complete set of features describing an instance, and it is more feasible for deep learning algorithms. They offer a large amount of logged data, allowing users to set their tasks better but with more data preprocessing. Organizers offered only the fittest in the EdNet competition site (http://ednet-leaderboard.s3-website-ap-northeast-1.amazonaws.com/) in the fittest sub-dataset (KT1), which consist of students' question-solving logs, and the best results achieved 0.7368 accuracies and 0.7811 for AUC.

Riiid! Answer Correctness Prediction[42] is a very new dataset available on the Kaggle platform and has a competition assigned to it. The e-Learning context consists of lectures and questions stored in two *.csv* files. These files contain information about 418 different lectures, mainly composed of 230 concepts and 170 solving questions. We mention that the questions are of multiple choice. The training file gathers recorded actions of 393,656 unique users that asked 13,523 specific questions. The public leaderboard gathers the scores from 1521 teams which is a clear indication regarding the quality of the dataset. The task associated with the dataset is to build a knowledge tracing model that predicts how students will perform on future interactions, and the validation uses the classical area under the ROC curve metric.

---

[39] KC Modeling for Programming, https://pslcdatashop.web.cmu.edu/Project?id=294.

[40] CSEDM 2020, https://sites.google.com/ncsu.edu/csedm-ws-edm-2020/data-challenge.

[41] EdNet dataset, https://github.com/riiid/ednet.

[42] Riiid AIEd Challenge 2020, https://www.kaggle.com/c/riiid-test-answer-prediction.

Multimodal learning Math Data Corpus[43] has been described in [38], a paper with 26 citations. A particularity of the dataset is that it contains high-fidelity time-synchronized multimodal data recordings (speech, digital pen, images) on collaborating groups of students as they work together to solve mathematics problems. The tackled issues are expertise estimation, prediction of problem-solving in mathematics or prediction of participation style.

Learn Moodle August 2016[44] is also a standalone dataset that has no published paper describing the dataset. Which is hardly used in cited research; therefore, its utility is rather challenging to be appraised. Still, the dataset has a detailed description as it gathers data from 6119 students exported in six *.csv* files. The most significant shortcoming is that the data files contain basic information, such that data analysis may require heavy preprocessing to obtain well-formed train and test data.

Lix Puzzle-game Data Set[45] is shortly described in [39] and contains 15 *.cvs* files, one for each participant at experiments of collecting data from gaming interactions while playing the game Lix. Each file contains 11 features describing in detail the actions performed during gameplay. Currently, there are only two citations, and game monitoring is not strictly related to educational processes.

Student Life Dataset[46] is described in [40]. Later [41], used the dataset to assess mental health and academic performance on the data that encapsulates the actions collected from 48 undergraduate students during a 10-week long period. The dataset is very well maintained and documented, having a list of 16 publications and eight presentations available on the site. The publications tend to analyze the dataset mainly from the psychological perspective, and one of the most exciting tasks is to predict the student's GPA based on the data collected from the phones. One valuable mention is that an R package (https://github.com/frycast/studentlife) is available for this dataset, which aims to help navigate and analyze. As the current number of citations is 577, we conclude that the dataset has been intensely used successfully in experiments.

Dataset for empirical evaluation of entry requirements into engineering undergraduate programs in a Nigerian University[47] is described in [42]. The dataset covers Engineering Education and gathers raw data from 2005 until 2009 and statistics regarding age and four scores with a target class. Among the tackled tasks from the nine citations, we mention predicting first-year student performance or analyzing the relationship between students' first-year results and final graduation grades.

MUTLA: A Large-Scale Dataset for Multimodal Teaching and Learning Analytics[48] dataset is described in [43]. The dataset is very well described and covers

---

[43] Multimodal learning Math Data Corpus, http://mla.ucsd.edu/data.

[44] Learn Moodle August 2016, https://research.moodle.org/158.

[45] Lix Puzzle-game, https://sites.google.com/site/learninganalyticsforall/data-sets/lix-dataset.

[46] Student Life Dataset, http://studentlife.cs.dartmouth.edu.

[47] Dataset for empirical evaluation of entry requirements, https://tinyurl.com/yxqf2v42.

[48] MUTLA, https://tinyurl.com/SAILdata.

many academic subjects (i.e., Mathematics, English, Physics, and Chemistry) and gather data from three sources: user records at question level log of student responses, brainwave data and webcam data. Although the dataset has not had any known usage or citation yet, the description and available data make it very suitable for EDM processes.

4 Proposed Methodology for Building Datasets

4.1 The Methodology Used for Data Collection

The data was gathered during the *Data Structures and Algorithms* course and is based on the graphs related topics. The Graph subjects represent half of the Data Structures course and are taught in the middle of the semester, so deepening this will significantly impact the exam grade. Another advantage is that if we can predict the student's failure during this period, there is still lots of time to catch up, so building a dataset and a system that can trigger a failure alert can be very useful for students taking this course.

The workflow for building the dataset starts from the student who takes tests; then these tests are recorded in the e-Learning platform, and after the tests period is finished, we can export the data saved in the database and feed them to the dataset generator tool. The tool can execute queries on the database, compute the attributes for each student, and construct the dataset, which can be utilized for triggering failure to the student, provide feedback regarding the knowledge level, or predict the final exam result.

The testing phase started after the graph's chapters were lectured and the student had enough experience to answer the questions. The testing method is based on a concept map representation which is a directed acyclic graph. Students had to take minimum five tests from the graphs to have a testing grade computed and contained in the dataset. The amount of questions for each test differs from eight to ten because we logged in this dataset two years of studies, and the first year we considered a number of ten questions for each test, and in the next year, we reduced the number of questions. The reason for reducing from ten to eight questions for each test is that students could not complete the questions from the last concepts of the graph in several cases.

Table 1 presents the concepts along with the number of questions assigned to each concept. There are 98 concepts allocated to 11 concepts, and their reliance graph is presented in Fig. 1. The sequence of the concepts in the graph matches to the succession during the semester, so first, they will discover Representations, then GSearch, etc. The initial test which will be taken will have questions from the *Representations* concept, and then and then after the student answer very good to the questions from this concept, he will proceed to GSearch. The edge for proceeding from a concept to another is 50% and to have a concept proficiency is 75%, so a student

Table 1 Concepts and questions distribution

Concept	No. of questions
Representations	15
GSearch	5
DFS	11
BFS	5
SSSP-fundamentals	8
Dijkstra	12
Bellman-Ford	8
APSP	10
SSSP/APSP-Wrapup	7
MST	9
Misc	8

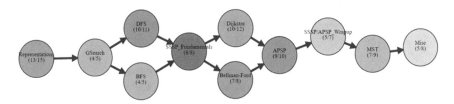

Fig. 1 Concept's graph example

that took tests from GSearch answered correctly to at least 50% of the questions from Representations and in order to skip getting questions from Representations he must have answered properly to more than 75%.

In this scenario, we aim to explore the dataset and examine if we can forecast the tests grade and the final exam based on the recorded data. A good relationship between attributes computed in the dataset and any of the class labels will make the attributes we computed good prognosticators for the exam or test results. The conduct experiment is performed using machine learning algorithms, considering regression and classification tasks. For the experiments we used the Weka machine learning library to experiment with this dataset, which offers an easy-to-use range of algorithms. For all the algorithms used in our researches, we chose the default values for the tuning parameters offered by Weka. There are two other kernels which can be used for broaden and more complex experiments on Kaggle.[49]

---

[49] https://www.kaggle.com/cristianmihaescu/dsa-test-dataset/kernels.

4.2 Structure of the Dataset

The dataset consists of eleven attributes that describe the student activity performed by 275 students (instances) performed during tests. We consider eleven features even if in the list in which we present them we count thirteen features because *MeanTestGrade* is the similar but in two forms: with continuous values and with discretized values and the exam grade also have two types: with exact values obtained at the exam and with binary values (0 and 1) which express failure or success. The features are relevant for the activity performed during the examination period, and the data was collected in two different years of study: 2018 and 2019. The dataset is publicly available on Kaggle; the features are presented below and can be used for further experiments.

1. *NumberOfLogins*—the number of logins on the platform made by the student during the testing period
2. *TimeSpentOnPlatform*—r5678 the whole time spent on the e-Learning platform
3. *NumberOfTests*—the complete number of tests the students did
4. *TimeSpentForTests*—the whole time spent by the students on platform tacking tests
5. · *AverageTimePerTest*—the mean time spent for taking a test
6. *NumberOfConcepts*—the full number of concepts included in student's tests
7. *NumberOfActions*—the complete number of actions logged for the student
8. *NumberOfRevisions*—the number of times a student revised the questions from past tests
9. *LastGrade*—the grade taken at his last test
10. *MeanTestsGrade*—the mean grade taken from all tests
11. *MeanTestsGradeD*—the same mean grade but discretized with in the range from 4 to 10
12. *ExamGradeD*—the finishing grade obtained by the student at the final exam, which is a discretized value in this case
13. *PassExam*—the finishing grade divided into binary values: 0 for failure and 1 for passing the exam.

Table 2 presents the attributes and a short analysis regarding their values. We focus on mean, minimum and maximum values along with standard deviation, which is included in the last column because most of the values are numeric, and these metrics are relevant in this case. Attribute 11 from the table has discrete values, and we cannot compute mean and standard deviation, so we have "*NA*" on those columns.

Each presented feature is essential for predicting the final grade and learner's engagement in the learning activity. The logins number and time spent on the platform and number of logged actions are influencing the grade as a quantity metric (number) on how much the learners are engaged in a learning activity through the platform as a more significant value implies more engagement. Usually, a more engaged learner will also be interested in gaining better grades and knowledge improving.

Table 2 Attributes and short statistics

Attribute	Min	Max	Mean	StdDev
NumberOfLogins	1	42	7.57	5.74
TimeSpentOnPlatform	2	948	188.95	160.18
NumberOfTests	1	28	8.5	5.39
TimeSpentForTests	0	122	37.44	23.47
AverageTimePerTest	0	9.1	4.14	1.7
NumberOfConcepts	0	11	9.96	2.62
NumberOfActions	5	330	74.67	49.6
NumberOfRevisions	0	72	12.25	10.46
LastGrade	1	10	6.67	2.39
MeanTestsGrade	1.45	9.77	6.41	1.7
MeanTestsGradeD	4	10	NA	NA
ExamGradeD	3	10	6.44	1.82

The number of tests is significant for the dataset and the final grade because it is a good indicator of how good the learner's implication is in the learning process and how significant the influence of the other actions on the final result is. Figure 2 presents the distribution of the tests, and on the OX axis, we have the number of the tests (from 0 to 28) taken, and on OY, we have the number of tests taken by a student. Analyzing the figure, we can see that most of the students took fewer than 14 tests, and the colours are blue for failure at the final exam and red for success.

The number of concepts addressed by the learner is relevant because they indicate how good the student's progress is. The relevance comes from the recommender system already implemented in the platform [44], which allows students to get questions related to a concept only after passing a certain threshold, which signifies that he knows previous concepts well.

The number of revisions refers mainly to how many times learners accessed a past test to see which questions were correctly answered and the correct answer for the

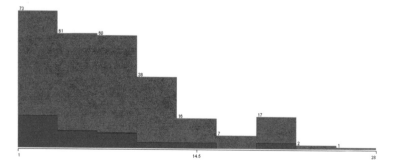

Fig. 2 Number of tests distribution

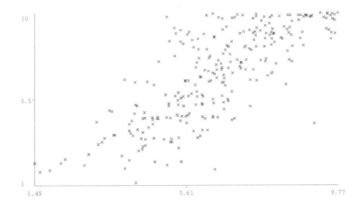

Fig. 3 MeanTestGrade versus LastGrade Correlation

questions. This concept is relevant for learning engagement and predicting the final grade because a higher interest in past tests means that the student aims for better results and/or wants to improve his knowledge. This feature, itself, can provide a significant insight regarding the learner's behaviour because there are several cases: have an excellent grade and revise the questions, have a significant grade but omit the revising and then the other two: have a low grade and revising and have a low grade but not revising the answers. These four situations also need to be addressed as future work as they provide valuable data analyzed can help the student's modelling and improve the grade or failure predicting.

The final grade is relevant for the final grade or the student's level of knowledge because it marks at what level the learner stopped taking tests, and another motivation for computing this attribute is the correlation between it and the final result. In Fig. 3, there is a correlation between *LastGrade* on the OY axis and *MeanTestGrade* on the OX axis, and it is visually evident that there is a correlation between these two attributes. The colour of the points from the figure corresponds to the final grade, and we chose blue for fail and red for passing.

The mean grade for tests is presented in two ways, the average grade computed from all the tests and a discretized version. The grade discretization process is made considering that if it is more significant than 0.5 points, we consider it the next grade, and what is less means the integer. For example, 4.3 will be considered as four, while 4.7 will be discretized as a five. The first (exact) version of the grade is more accurate than discretized because the note's value was not approximated, but it limits us to mainly regression algorithms, but discretized version allows us to use a wider variety of algorithms like the classification algorithms.

Exam grade was already discretized, and it is the grade obtained by the student at the final exam. Predicting the exam grade based on the previous features, including the MeanTestGrade, is an excellent way to predict the student's failure. The Mean-TestGrade can also be a robust estimator for exam failure, and we aim to prevent it by making recommendations to students based on their testing results.

Fig. 4 Final grade distribution

Fig. 5 Final grade discretized into two attributes

The motivation for adding the *PassExam* attribute can be deducted from comparing Fig. 4 with Fig. 5. In Fig. 4, we have the grade distribution, so we have seven classes from which one is for failing the exam, and the rest of seven is for passing; blue colour corresponds to failing and red for passing. Dividing the data into two classes, like the ones presented in Fig. 5, reduces the information gain and reduces the problem to binary classification even if we have to deal with imbalanced classes (Figs. 4 and 5).

5 Conclusions

This chapter tackles the existing educational datasets focusing on education, offering an overview of the current state of the art datasets. This datasets overview is relevant for the Educational Data Mining research area, making it easier to pick a suitable dataset for a specific problem. There is a wide variety of datasets that are more or less known in the community, and several of them can be used for a significant variety of tasks, and some of them are very focused on a specific problem. Knowing the state of the domain at this moment is relevant not only for choosing the suitable dataset but also for extending or extracting new datasets. Another problem that is still open in this area is the results and algorithms used for solving specific tasks because many datasets already have notebooks, scrips and tools, some of which do not, and in some cases, there is also plenty of space for improvement.

We also focused on one specific dataset—the DSA course dataset, which was contributed to Kaggle and included specific characteristics like generic features, friendly format, and is ready to use without any preprocessing. These characteristics allow data scientists to quickly experiment and model plenty of problems by eliminating the less relevant attributes. There is still a lot of work in this area as other methods besides the ones presented in the paper should be explored to achieve superior performance.

References

1. Popescu, P.S., Mihaescu, M.C., Teodorescu, O.M., Mocanu, M.: Student testing activity dataset from data structures course. In: RoCHI-International Conference on Human-Computer Interaction, p. 157 (2020)
2. Mihaescu, M.C., Popescu, P.S.: Review on publicly available datasets for educational data mining. Wiley Interdiscip. Rev.: Data Min. Knowl. Discov. 11(3), e1403 (2021)
3. Romero, C., Ventura, S.: Educational data mining: a survey from 1995 to 2005. Expert Syst. Appl. 33(1), 135–146 (2007)
4. Baker, R.S., Yacef, K.: The state of educational data mining in 2009: a review and future visions. J. Educ. Data Min. 1(1), 3–17 (2009)
5. Romero, C., Ventura, S.: Educational data mining: a review of the state of the art. IEEE Trans. Syst. Man Cybern. Part C (Appl. Rev.) 40(6), 601–618 (2010)
6. Koedinger, K.R., Baker, R.S., Cunningham, K., Skogsholm, A., Leber, B., Stamper, J.: A data repository for the EDM community: the PSLC DataShop. Handb. Educ. Data Min. 43, 43–56 (2010)
7. Merceron, A.: Educational data mining/learning analytics: methods, tasks and current trends. In: DeLFI Workshops, pp. 101–109 (2015)
8. Dutt, A., Ismail, M.A., Herawan, T.: A systematic review on educational data mining. IEEE Access 5, 15991–16005 (2017)
9. Silva, C., Fonseca, J.: Educational Data Mining: a literature review. In: Europe and MENA Cooperation Advances in Information and Communication Technologies, pp. 87–94 (2017)
10. Rodrigues, M.W., Isotani, S., Zarate, L.E.: Educational data mining: a review of evaluation process in the e-learning. Telemat. Inform. 35(6), 1701–1717 (2018)
11. Romero, C., Ventura, S.: Data mining in education. Wiley Interdiscip. Rev.: Data Min. Knowl. Discov. 3(1), 12–27 (2013)
12. Romero, C., Ventura, S.: Educational data science in massive open online courses. Wiley Interdiscip. Rev.: Data Min. Knowl. Discov. 7(1), e1187 (2017)
13. Romero, C., Ventura, S.: Educational data mining and learning analytics: an updated survey. Wiley Interdiscip. Rev.: Data Min. Knowl. Discov. 10(3), e1355 (2020)
14. Slater, S., Joksimović, S., Kovanovic, V., Baker, R.S., Gasevic, D.: Tools for educational data mining: a review. J. Educ. Behav. Stat. 42(1), 85–106 (2017)
15. Ihantola, P., Vihavainen, A., Ahadi, A., Butler, M., Börstler, J., Edwards, S.H., Toll, D.: Educational data mining and learning analytics in programming: Literature review and case studies. In: Proceedings of the 2015 ITiCSE on Working Group Reports, pp. 41–63 (2015)
16. Alon, U., Zilberstein, M., Levy, O., Yahav, E.: code2vec: learning distributed representations of code. Proc. ACM Programm. Lang. 3(POPL), 1–29 (2019)
17. Vieira, C., Parsons, P., Byrd, V.: Visual learning analytics of educational data: a systematic literature review and research agenda. Comput. Educ. 122, 119–135 (2018)
18. Ferreira-Mello, R., André, M., Pinheiro, A., Costa, E., Romero, C.: Text mining in education. Wiley Interdiscip. Rev.: Data Min. Knowl. Discov. 9(6), e1332 (2019)
19. Bakhshinategh, B., Zaiane, O.R., ElAtia, S., Ipperciel, D.: Educational data mining applications and tasks: a survey of the last 10 years. Educ. Inf. Technol. 23(1), 537–553 (2018)
20. Loh, W.Y., Shih, Y.S.: Split selection methods for classification trees. Stat. Sin. 815–840 (1997)

21. Cortez, P., Silva, A.M.G.: Using data mining to predict secondary school student performance (2008)
22. Kahraman, H.T., Sagiroglu, S., Colak, I.: The development of intuitive knowledge classifier and the modeling of domain dependent data. Knowl.-Based Syst. **37**, 283–295 (2013)
23. Vahdat, M., Oneto, L., Anguita, D., Funk, M., Rauterberg, M.: A learning analytics approach to correlate the academic achievements of students with interaction data from an educational simulator. In: European Conference on Technology Enhanced Learning, pp. 352–366. Springer, Cham (2015)
24. Kuzilek, J., Hlosta, M., Zdrahal, Z.: Open university learning analytics dataset. Sci. Data **4**(1), 1–8 (2017)
25. Hussain, S., Dahan, N.A., Ba-Alwib, F.M., Ribata, N.: Educational data mining and analysis of students' academic performance using WEKA. Indones. J. Electr. Eng. Comput. Sci. **9**(2), 447–459 (2018)
26. Bhoi, N.K.: Mendeley data repository as a platform for research data management. Marching Libr.: Manag. Ski. Technol. Competencies, 481–487 (2018)
27. Kastrati, Z., Kurti, A., Imran, A.S.: WET: Word embedding-topic distribution vectors for MOOC video lectures dataset. Data Brief **28**, 105090 (2020)
28. Gómez-Tejedor, J.A., Vidaurre, A., Tort-Ausina, I., Mateo, J.M., Serrano, M.A., Meseguer-Dueñas, J.M., et al.: Data set on the effectiveness of flip teaching on engineering students' performance in the physics lab compared to traditional methodology. Data Brief **28**, 104915 (2020)
29. Prasojo, L.D., Habibi, A., Yaakob, M.F.M., Pratama, R., Yusof, M.R., Mukminin, A., Hanum, F.: Teachers' burnout: A SEM analysis in an Asian context. Heliyon **6**(1), e03144 (2020)
30. Delahoz-Dominguez, E., Zuluaga, R., Fontalvo-Herrera, T.: Dataset of academic performance evolution for engineering students. Data Brief **30**, 105537 (2020)
31. Hou, Y., Li, L., Li, B., Liu, J.: An anti-noise ensemble algorithm for imbalance classification. Intell. Data Anal. **23**(6), 1205–1217 (2019)
32. Ho, A., Reich, J., Nesterko, S., Seaton, D., Mullaney, T., Waldo, J., Chuang, I.: HarvardX and MITx: the first year of open online courses, fall 2012-summer 2013 (HarvardX and MITx Working Paper No. 1) (2014)
33. Kellogg, S., Edelmann, A.: Massively open online course for educators (MOOC-E d) network dataset. Br. J. Educ. Technol. **46**(5), 977–983 (2015)
34. Northcutt, C.G., Ho, A.D., Chuang, I.L.: Detecting and preventing "multiple-account" cheating in massive open online courses. Comput. Educ. **100**, 71–80 (2016)
35. Stamper, J., Niculescu-Mizil, A., Ritter, S., Gordon, G.J., Koedinger, K.R.: Bridge to algebra 2006–2007. Development data set from KDD cup 2010 educational data mining challenge (2010)
36. Amrieh, E.A., Hamtini, T., Aljarah, I.: Mining educational data to predict student's academic performance using ensemble methods. Int. J. Database Theory Appl. **9**(8), 119–136 (2016)
37. Choi, Y., Lee, Y., Shin, D., Cho, J., Park, S., Lee, S., et al.: Ednet: a large-scale hierarchical dataset in education. In: International Conference on Artificial Intelligence in Education, pp. 69–73. Springer, Cham (2020)
38. Oviatt, S., Cohen, A., Weibel, N.: Multimodal learning analytics: description of math data corpus for ICMI grand challenge workshop. In: Proceedings of the 15th ACM on International Conference on Multimodal Interaction, pp. 563–568 (2013)
39. Vahdat, M., Carvalho, M.B., Funk, M., Rauterberg, M., Hu, J., Anguita, D. Learning analytics for a puzzle game to discover the puzzle-solving tactics of players. In: European Conference on Technology Enhanced Learning, pp. 673–677. Springer, Cham (2016)
40. Wang, R., Chen, F., Chen, Z., Li, T., Harari, G., Tignor, S., et al.: StudentLife: assessing mental health, academic performance and behavioral trends of college students using smartphones. In: Proceedings of the 2014 ACM International Joint Conference on Pervasive and Ubiquitous Computing, pp. 3–14 (2014)
41. Wang, R., Chen, F., Chen, Z., Li, T., Harari, G., Tignor, S., et al.: StudentLife: Using smartphones to assess mental health and academic performance of college students. In: Mobile Health, pp. 7–33. Springer, Cham (2017)

42. Odukoya, J.A., Popoola, S.I., Atayero, A.A., Omole, D.O., Badejo, J.A., John, T.M., Olowo, O.O.: Learning analytics: dataset for empirical evaluation of entry requirements into engineering undergraduate programs in a Nigerian university. Data Brief **17**, 998–1014 (2018)
43. Xu, F., Wu, L., Thai, K.P., Hsu, C., Wang, W., Tong, R.: MUTLA: a large-scale dataset for multimodal teaching and learning analytics. arXiv:1910.06078 (2019)
44. Teodorescu, O.M., Popescu, P.S., Mihaescu, M.C.: Taking e-assessment quizzes-a case study with an SVD based recommender system. In: International Conference on Intelligent Data Engineering and Automated Learning, pp. 829–837. Springer, Cham (2018)

Building Data Analysis Workflows that Provide Personalized Recommendations for Students

M. C. Mihăescu⊙, P. S. Popescu, and M. L. Mocanu

Abstract Providing personalized recommendations in e-Learning systems represents one of the key aspects that has been enabled by the advances in Machine Learning. The ingredients that may be used to address this challenge reduce to building custom data analysis workflows that have input the context of the e-Learning system and output relevant recommendations as message or learning resources that need to be studied. This chapter presents two workflows that address two aspects that represent the main business logic within the pipeline: course difficulty ranking and message recommendations. The custom-designed mathematical modelling of the course difficulty ranking and the workflow of the feedback system are presented in detail. Experimental results show that the proposed ranking method and the decision tree-based model's accuracy perform reasonably well. Further works regarding the structure of the data analysis pipeline in terms of preprocessing, integration of state-of-the-art classification models, and other validation methods may further improve the relevance of the recommendations and open the way towards a more effective e-Learning system.

Keywords Data analysis · Machine learning · Classification · Ranking · Recommendations

1 Introduction

Building personalized recommendations for students is critical in any e-Learning system that heavily relies on a custom-designed data analysis pipeline. Within the designed workflow, there are usually modules that implement specific functionalities such as course difficulty estimator, ranking the difficulty level of knowledge units, usage of machine learning models (i.e., decision tree classifiers, SVM, etc.) to determine recommended messages or the classical problem of predicting student's performance.

M. C. Mihăescu (✉) · P. S. Popescu · M. L. Mocanu
Department of Computer Science and Information Technology, University of Craiova, Str. A.I. Cuza, Nr. 13, Craiova, Romania
e-mail: cristian.mihaescu@edu.ucv.ro

© The Author(s), under exclusive license to Springer Nature Switzerland AG 2022   43
M. Mihaescu and M. Mihaescu (eds.), *Data Analytics in e-Learning: Approaches and Applications*, Intelligent Systems Reference Library 220,
https://doi.org/10.1007/978-3-030-96644-7_3

Regarding the course difficulty estimation [1] tried to estimate the learners' subjective impression regarding the difficulty of course materials. The authors used the SVM algorithm and attempted to model students based on their head movements captured with a complex system attached to the monitor. Experimental results show that the average accuracy of estimations was 85,8%. Still, the limitation is that professors can see how challenging a course is, but they do not know how it should be adjusted.

Later, Liu [2] addressed the problem of raking the difficulty level of knowledge units based on learning dependency by assigning difficulty indicators to the knowledge units. The approach related to learning dependency implies a significant amount of subjectivity because certified persons must rank knowledge units with relevant knowledge in the area. The target within our proposed approach regards eliminating the human factor and rely only on data-driven analysis. We design the ranking procedure based on the number of critical concepts introduced by each course and chapter instead of depending on the human factor.

The task of intelligent data analysis in education environments starts 1995 as presented in a review paper [3], but course and students evaluation is addressed by [4–7]. Among the most critical factors that drive a successful e-Learning platform [8] proposed a questionnaire to understand essential factors that affect learners' satisfaction in e-Learning environments. The results exposed that professors' attitudes toward the online educational environment, learner computer anxiety, course flexibility, course quality, perceived ease of use, usefulness, and diversity in evaluations and homework are critical aspects that affect learners' apparent satisfaction.

One of the first approaches regarding the messages recommendation procedure is Abel [9] that build ontologies mined from Comtella-D, a "discussion for a" used for learning purposes. The authors describe a high-level system but with any synthetic relevant results. Their approach was using ontologies and SILK for the recommendation system. In this case, the preprocessing step is very complicated, and the result is given based on classes, as many other systems do. The system's drawbacks appear when users start a discussion in the wrong thread, or one topic fits in multiple threads. Also, the recommendation based on classes is better than the recommendation based on leaves as the proposed approach may be further refined.

Our proposed approach refers to analyzing the messaging interaction between users for identifying critical learners. Providing feedback regarding messaging activity and putting into correspondence several users create a semi-structured data format. Critical learners are essential because they might be trend makers, opinion makers or just famous students. In any of the already mentioned cases, those learners need to be identified and worked upon in an attend to improve the learners' knowledge level. Evaluation of the proposed system represents a critical issue as with all recommender systems of any type.

Most proposed messaging recommender systems try to find any pattern that can be recognized in the logged data that might predict a learner's final result. We try to find a pattern in the interactions students have with colleagues or professors, which can predict the final result. Another addressed research issue refers to providing feedback about the messaging activity. This feedback can be precious if we think about the recommender systems and may improve social skills.

Predicting students' performances [10] is one of the critical ingredients addressed by researchers and integraintod in many workflows. In Davies [11], authors try to find a correlation between students' online interactions and learning performance. The study was performed on 122 undergraduates, which examined the frequency of their online interactions. A comparison of frequencies with grades was performed at the end of the year. The results revealed that a higher online interaction did not lead to a significant increase in performance; however, it was noted that students who failed in their courses tended to interact less frequently. At about the same time, Wang [12] presented a study conducted on graduate-level online courses by examining the relationship between student visibility and learning outcomes. The visibility in this study refers to emotive [13, 14] social and cognitive presence.

Regarding the algorithm's performances in PURDILĂ [15], a fast decision tree algorithm is presented, more precisely, a fast pruning algorithm. This algorithm uses a compression mechanism to store the training and test dataset in memory.

More recent research [16] aims to analyze learners' performance by using a data mining approach, usage of classification (i.e., decision trees) models or clustering to predict the learners' final results. This approach is fundamental because a significant amount of information can be lost using only class analysis. The authors state that using decision trees combined with Gini index can have a better prediction accuracy. Still, in most cases, the accuracy obtained after cross-validation is the only measure that can give us an overview of how well a classifier performs. The results presented in the paper refer only to the classification accuracy, and even so, the average value is around 70% which is not very good.

Further, Daud [17] aims to predict students' performance using advanced learning analytics and the authors also present visualization techniques. The experimental results are done using Weka 3.7, but the dataset has only 100 instances and 23 features. The feature ratio regarding the instances is undoubtedly not the best. The performance evaluation is performed using precision, recall, and F1 score despite computing accuracy like in most cases. One drawback is that the construction of the dataset might be biased because of the selection of the students (50% completed, 50% dropped), and their approach using above mentioned precision metrics does not offer noticeable results.

One interesting method of improving the classifier's performance is presented in Iwata [18], in which the authors propose a framework for improving the classifier's performance by effectively using auxiliary samples. The auxiliary samples are labelled not regarding the target taxonomy according to which we wish to classify samples but according to classification schemes or taxonomies different from the target taxonomy. Their method proposes a classifier by minimizing a weighted error over the target and auxiliary samples.

Another technique for improving the classifier's accuracy [19] is the use of attribute bagging and boosting [20]. In Bryll [21], the authors present how using the attribute bagging technique improves both the accuracy and the reliability of classifier ensembles; this is induced by using random subsets of attributes. Attribute bagging is a generic method that can be used with any machine learning algorithm. The presented approach establishes a custom subset of attributes, and then it randomly selects several subsets of features, creating projections of the training set on which the ensemble classifiers are built. The induced classifiers are then used for voting. That article compares the performance of their AB method with bagging and other algorithms on a hand-pose recognition dataset. It is presented that the AB approach offers consistently improved outcomes than bagging, both in terms of stability and accuracy.

An early paper by Minaei-Bidgoli [22] presents an exciting approach that aims to classify students to predict their final grade. The features were extracted from the data logged in an education Web-based system. The authors design, implement, and evaluate many pattern classifiers and perform a benchmark comparison on their performance on a dataset obtained from an online course.

2 Machine Learning Workflows

Machine learning workflows—especially in classification contexts—use a dataset with labelled instances for training. Then, based on what the algorithm learns from that dataset, it will label unseen data. In some cases, we call this a prediction because it is based on past events. Therefore, having a new instance with all its features but with no given value for the target feature, we can further use the model to predict the result (i.e., the value of the target feature). It is essential to have the same features on the learning data and the test or predict data.

One crucial step that influences the process of learning is data acquisition and preprocessing. The data gathering process usually refers to Data Mining (DM) and the analysis and examination process. This step determines how many features an instance will have and, more importantly, what features will have. There is a trade-off when choosing the features that will describe the data because choosing too many features to describe an instance may need a large amount of data (many instances) or may produce low accuracy. Losing or just eliminating some features may provide overfitting or false positives. Once the data mining process finishes, the most significant step is choosing the proper algorithm.

The data used for machine learning algorithms can be gathered from any application domain and have any number of numeric or nominal type features. There are two main rules if we desire a significant result: *the features that describe the data needs to be significant for the classification task, and the data must be large enough to produce an unbiased result.*

A dataset is usually a set of information that was previously preprocessed and have a standard format. These datasets are saved mainly in ".*csv*" and ".*arff*" formats depending on the library, or the tool was used and needed to have at least two components: *the name of the features and the instances.*

In many cases, after choosing the feature's names, there are specific datatypes or range of values defined for them. Depending on the desired task, a class feature will be there and depending on the type of the dataset, some features may have a value or not. Datasets are mainly of three types: *Training datasets, testing datasets, and new/unseen datasets.*

In order to create the models, we need a well-constructed dataset; it is necessary to have labelled data if it is a supervised learning algorithm, and if we talk about unsupervised learning, there is no need to have labelled data. Depending on the chosen algorithm evaluation technique, there may be or not a separated testing dataset. If the attributes from the dataset fed the supervised learning algorithm got labels, the test dataset must have values for the labels (or classes). The new datasets are datasets with unlabeled data; the class feature exists, but there will be no data for it; after using the previously constructed model, the class feature will receive values.

Machine learning algorithms separate into two main classes: algorithms that use supervised learning and algorithms that use unsupervised learning. In this chapter, we will refer only to the supervised learning algorithms. Unsupervised learning refers mainly to clustering algorithms that work to make instances based on their standard features. Supervised learning algorithms need two different datasets, one for the training step and one for the test step; unsupervised learning does just clusters of items from the training data. Every algorithm has specific capabilities and is suitable for specific tasks. Moreover, each algorithm also has several tuning parameters.

Figure 1 presents the main steps regarding the process of employing an ML algorithm. Usually, the data gathered needs to be processed to meet the requirements to be used as a training or test dataset. After we have a valid dataset, an ML algorithm can be used to produce a model. This step usually includes choosing the algorithm and tuning the algorithm to get the best accuracy on the training data. Depending on the data we got, the algorithm's evaluation can be made on the training data or

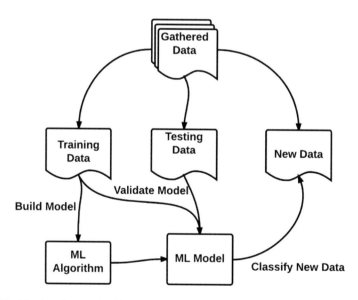

Fig. 1 Machine learning running flow

specific testing data; further discussions will be on this problem. Once we obtain the ML model and predict accuracy is good enough, we can classify new data. All three datasets from the figure will have the same structure and features. If some features are omitted from the training dataset during the data tuning process, we need to make the same changes on the other datasets. A refined version of the workflow that creates a system for course evaluation is presented in Fig. 2.

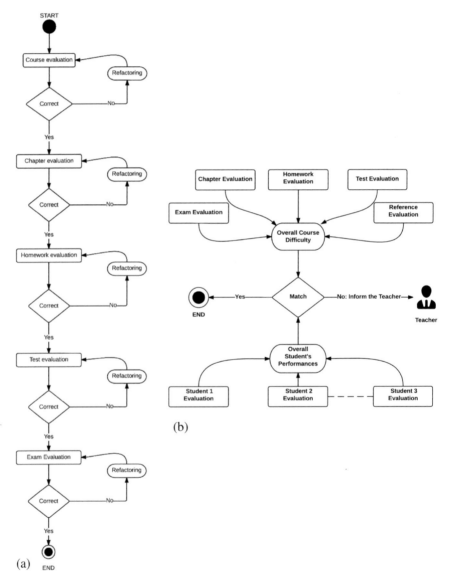

Fig. 2 a) Course evaluation flow; **b)** System overview

We refer only to classification algorithms in this context, and the datasets used were from the educational research area. Depending on the research area on which we use machine learning algorithms, specific tasks need to be performed, and there are specific datasets on which perform better some of the algorithms.

3 Inferring Personalized Recommendations by Course Difficulty Prediction and Ranking

One of the critical aspects of providing relevant recommendations is understanding the gap between student's level and course difficulty [23]. The end goal regards students as beneficiaries as they will get helpful learning resources and therefore have a better learning experience.

Regarding the recommendations for professors, one of the most significant problems is that they need to adjust the course materials for student's level and provide a uniform learning experience during the whole semester. Usually, learning resources have different difficulties levels because their concepts may be new or already known. Many new concepts regarding discipline will induce a high level of difficulty, and based on this assumption; we try to evaluate the difficulty of a course chapter to provide recommendations regarding the difficulty and how it can be adjusted. It is essential to mention that this approach matches the Tesys e-Learning platform [24] architecture entirely, but the learning resources referred to in this section are available in most e-Learning platforms.

Figure 2a) presents a little sequence diagram of the evaluation procedure. We see here the main steps that need to be considered when we must evaluate a course considering all its components.

First, the system evaluates the course material, which is a repository of chapters, to get an intuition regarding the overall difficulty and have a reference. After that, the evaluation of each chapter comes, and the goal is to have an overall well distributed difficulty level, and each chapter's level should be acceptable for each student. The first step represents an essential task because if we consider at least 12 chapters (for our learning programs, we have 14 weeks), some of them can have a different level of difficulty but may not influence the overall difficulty. Then, after there is suitable course material, we can go further at the homework evaluation. This context represents a different type of evaluation because of its purpose—student's knowledge evaluation. This feature allows us to have a different evaluation approach to validate the estimation after every deadline. After homework evaluation, we can evaluate the tests and finally the exam's difficulty.

Another module of the evaluation system is the one that aims to evaluate students in order to have a match between students and course difficulty. The students' evaluation module logs the data from the platform and then estimates the students' overall knowledge or specific knowledge. For this purpose, we need to analyze the mean of the difficulty against all the students and whether there is a match. Figure 2b) offers the system's overview and how the modules are distributed.

For the learning assets evaluation procedure, we propose a custom-designed work-flow. We analyze the whole set of learning assets based on the tests and exam results, but there can be differences from year to year. To be more exact, in year X, we may have some good students who will perform exceptionally well at the tests and exams, but in the next year $(X + 1)$, we can have students who find the courses very hard. Based on this assumption, the teachers are supposed to make the learning assets simpler for the year $(X + 2)$; but again, in the year $(X + 2)$, we may have some students better or as good as the students in the first year (i.e., year X). This context may yield an undesired situation because the students and the learning assets are opposite. To have a standard evaluation metric, we propose one based on terms extraction.

In our proposed approach, we introduce a method that processes the content of each document and determines the concepts that were explained in previous courses and the ones that were presented for the first time. If each new concept requires the same volume of effort to be assimilated by the student, we can establish a base-line effort of the material by calculating the number of new concepts included. We must also consider that each new concept must be understood in relation to the other concepts included in the material, adding more complexity to the overall difficulty level. Of course, the effort of each new concept is a factor to be taken into considera-tion, but that cannot be rigorously determined automatically so that it will be omitted. Considering this rationale, we can define D as the asset difficulty with the following formula:

$$D = \text{NC} \times C \tag{1}$$

where D represents the asset difficulty, NC represents the number of new concepts, and C represents the number of concepts from the learning asset.

As previously mentioned, the difficulty level of the course increases proportionally with the number of new concepts and the number of relations between those and the existing ones. For example, an online course that contains five known concepts and two new notions will have a difficulty level of 10, more significant than another one containing the same number of new concepts but fewer familiar concepts associated. Similarly, we can differentiate between the difficulty levels of two assets containing the same number of concepts that the student already understands but with different new notions.

For further clarifications, we define a set of concepts that are unique to the chapter:

$$N(x) = C(x) - C(x - 1) - C(x - 2) - \cdots - C(1) \tag{2}$$

where $C(x)$ is the set of concepts referred to in the analyzed chapter, $C(x - 1)$ represents the collection of concepts presented in the previous chapter, etc. The minus signs do not indicate in this situation arithmetic operations but subsequent subtractions between sets. Variables that are used in formula (2) represents the cardinality of the set $N(x)$:

$$NC = |N| \tag{3}$$

Each set of concepts is obtained automatically using a natural language processor, applying the classical TF–IDF technique and selecting the concepts with the highest weights.

The presented method may be applied to every learning asset. Every homework or question that references many concepts or few recently learned concepts could be labelled as hard. The number of concepts covered by the evaluation techniques can mark the level of difficulty for tests and exams.

The overall learning assets difficulty can be computed based on each learning asset difficulty. Based on this assumption, we derive the formula:

$$OD = \frac{\sum_i^n L_i}{n}. \tag{4}$$

where OD = overall difficulty and L_i is the difficulty computed for ith learning asset. For validation purposes, we choose a course, and we select the third chapter of the course. We perform a concept retrieval on the chapter, and then we compare obtained result with the concepts already extracted from the previous chapter. At this step we are able to compute the value of D for three new concepts from a total eleven concepts at respective chapter we obtain from formula (1) that $D = NC \times C \Rightarrow D = 3 \times 11 \Rightarrow D = 33$ as a numerical quantity describing the learning asset's difficulty.

Further, we compute NC (i.e., new concepts) by using the formula

$$NC = C(x) - C(x-1) - C(x-2) - \cdots - C(x-n) \tag{5}$$

And obtain that the number of new concepts is NC $= 12 - 4 - 5 \Rightarrow$ **NC** $= 3$.

Reiterating this step for each chapter of the discipline we can compute the mean difficulty of the course. Then we can compare it with the other learning asset that belongs to the selected discipline.

Using formula 4 we can determine the overall difficulty as

$$OD = \frac{24 + 30 + 33 + 28 + 35 + 37 + 23 + 31 + 30 + 35}{11} \Rightarrow \mathbf{OD} \simeq 30$$

Therefore, we have derived a practical approach that aims of determining the course difficulty (i.e., the difficulty of each learning asset that makes up the course), thus opening the way providing feedback for professors regarding the difficulty of their learning assets and providing students with study materials that are a good fit within the learning path.

The first aspect (i.e., determining the course difficulty) answers a question professors usually have in face-to-face and e-learning education: *"How difficult is the course? How hard are the learning assets that build up the course?"* Of course, answering these questions should be done regardless of student's knowledge level. Therefore recommendations in this matter should not depend specifically on the number of students or their interactions with the system. Still, once we have available the activity patterns of the students, these may be integrated with the estimated difficulty of learning assets, and therefore personalized recommendations may be inferred.

4 Personalized Message Recommendation by Usage of Decision Trees

Decision trees represent an essential part of classification algorithms, and the models constructed using them differ because decision trees can produce comprehensible and interpretable models, not just models that behave like black boxes. Several algorithms build decision trees and create classification models going from simple models like *DecisionStumps* to ensemble algorithms which usually combines many models to provide a high accuracy output. Even when using only the decision trees, there is a trade-off between interpretable and top accuracy models because ensembles built on simpler models may become very hard to interpret. Algorithms like J48 or ID3 can provide easy to understand models, but for several attributes' configurations, the model can be very noisy even in a pruned version.

Usually, a decision tree model has three types of nodes: root, intern nodes and leaves. The root usually marks the most significant attribute from the dataset, which splits the information gain into two comparable groups, and the significance of the subsequent splits gets lower as we go deeper in the tree. The internal nodes contribute with their splits to the rule that generates the leaf, in some cases, the rule can be simple, but in others, the rule that generates a leaf can be hard to interpret. On the last level of the tree, we have the leaves, which can be loaded in some cases with instances. A specific feature of the decision trees is that each leaf belongs to a class, but one class may have more than one leaf. The leaf analysis—presented in more detail in this chapter—revealed that even if two or more leaves belong to the same class, they will have several differences. We can perform this analysis only on particular decision trees models as many leaves may lead to a hardly understandable model, so two main ingredients lead to interpretable models: the algorithm and the dataset's structure.

J48 is C4.5 Ross Quinlan's algorithm [25] implementation in Weka open-source machine learning library [26], and this algorithm is used in most of the experiments presented in this thesis. The C4.5 algorithm is an improvement of the ID3 algorithm (Iterative Dichotomiser 3), which was implemented seven years earlier by the same author. Both algorithms have two steps: tree growth and pruning, but in the case of C4.5, in the pruning phase, internal nodes are replaced with the leaves. C4.5 is even more permissive than ID3 being able to manipulate both continuous and categorical attributes; this advantage makes the algorithm more flexible and allows us to feed it more variate data. C5.0, which is the next iteration of R. Quinlan's algorithm, is rarely included in open-source libraries, and the only advantages over C4.5 are the more scalable architecture.

C4.5 models are built iteratively, starting from the root and splitting the dataset at each step based on the information gain gathered from the attributes. The algorithms compute entropy iteratively for the remaining data, so we will have successive splits until each leaf encapsulates only the instances belonging to the same class. The results of applying a J48/C4.5 algorithm from Weka on a dataset is an interpretable decision tree model along with several evaluation metrics.

The advantages of using the J48/C4.5 algorithm on educational data are related to fast response and interpretable models. Most iterative algorithms that offer a model composed of one tree are reasonably interpretable, but not all provide high flexibility regarding the format of the features. Building just one model is usually fast for educational data which do not use large datasets, and the classification of a new item based on an existing model is most likely to be instant. There is a trade-off between straightforward J48/C4.5 models and ensembles is that the former provide better accuracy but more extensive build time. On the interpretability side, simpler models may offer concise rules from root to leaves, but ensembles tend to build more than one tree while improving the existing tree iteratively without knowing when the actual model is overfitted on the training dataset.

We further present an advanced messaging system that classifies users and gives them discussion fragments or full messages gathered from discussion forums [27, 28]. Considering the task mentioned above, the recommender system is based on classification and offers recommendations for students based on their messages. The educational environment is Tesys, an online educational platform where students, professors, and secretaries perform their responsibilities. Some of the current actions in this e-Learning platform are taking tests and exams, downloading courses, and communicating with professors. All performed activity represents the critical ingredient for the data analysis process because we can gather the necessary data needed for training the classification algorithm from this platform.

The system consists of two modules: the server-side platform and the client. On the server-side of the software system, we configure and run the classification process, gather data for the training set and generate the classification model. On the server-side, there are implemented three modules: the data gathering module, the model building module, the message indexing module, and the student classification and message retrieval module.

The first module will gather data from the database and the log files to generate the training files containing the values to all features that describe students. The model building module is responsible for creating the decision tree that has the features in its internal nodes and the class labels in leaf nodes. The message indexing module inserts messages into the index according to the student's class who wrote the message. The whole message is composed of the student's question along with the

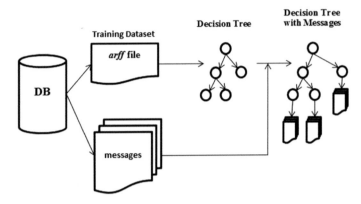

Fig. 3 Main workflow for indexing messages

professor's answer. The message retrieval module classifies the student and retrieves the associated messages with the obtained class label.

From the developer's point of view, when a student sends a message to a professor and the professor answers accordingly, the message will be indexed along with the student's class. Once the message is indexed, it will be available upon request to students with the same class label from the classifier's point of view.

Figure 3 presents the main data pipeline; the training set builder will create a *.arff* file with the input dataset based only on activities performed by students. After the *.arff* file is generated, it will remain on the system repository if a better challenger model is not available. When more data and more features for describing students are available, a challenger model may be created, and the old one may be replaced only if better classification accuracy is obtained.

The process of obtaining messages recommended for an individual student begins with the classification of the student interested in obtaining messages.

A visual description of the process regarding the message's retrieval is presented in Fig. 4. Once the student's class is determined, the associated messages to that class may be delivered as recommendations. Those messages are the ones that were sent by students that were pretty much in the same educational situation, and they may give him clarifications regarding several issues.

For the classification module, we used a data set gathered from the student's activity repository. The raw data is gathered from the Tesys database and then is converted into a .arff file. The .arff file will contain the features used in the classification process, which is listed in the table.

All features in this process are critical for the classification process due to their independence and are presented in Table 1. The total number of messages, number of tests, the time spent online, and the average mark obtained represent the status of a student when sending a message to a professor.

The first feature, *noOfMessages* ensures us the student's interest in using the system and his communication skills. In the *arff* file, this feature's type is numeric, and its value equals the sum of the sent and received messages.

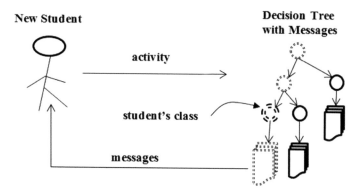

Fig. 4 Main workflow regarding messages retrieval

Table 1 Description of the features used for the student classification	Feature name	Description
	noOfMessages	The number of messages sent and received on the platform
	noOfTests	The number of performed tests
	timeOnLine	The time spent online having activity
	avgTestMark	The average of the results obtained in the tests
	Class	The target class to which the student belongs

The feature called *"noOfTests"* represents the number of tests taken by a student. It has a numeric type, the same as *timeOnLine* and gives a better relevance for *avgTest-Mark*. This attribute signifies the average mark obtained in tests and is discretized, resulting in three values: low, average, and high, depending on his results.

The target class attribute is related to the student's final mark and is discretized in *weak*, *average*, *good* and *excellent* depending on his mark.

In this system, we have included a feedback mechanism that will improve the classification algorithm's performance. Every time a student receives a message from the system, he can mark the message as relevant or not to him. Once the feedback is received, the message's weight is increased. This mechanism gives a more substantial weight value to inspiring messages, and thus future students will obtain more relevant feedback.

Usage of this approach, the system will continuously increase its relevancy in recommended messages. A better representation of the feedback system is presented below in Fig. 5.

A short example of the training set is presented below in Fig. 6.

For prototype testing purposes, we considered two short tests using synthetic data, and we found that the number of correctly classified instances increases along with the number of instances in the *.arff* file. Figure 7 presents the snipping of the

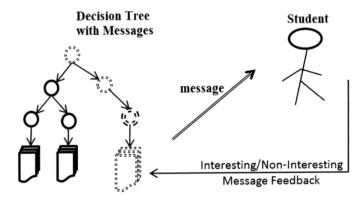

Fig. 5 Data workflow for the feedback system

```
@RELATION TrainStud

@ATTRIBUTE nrOfMessages numeric
@ATTRIBUTE nrOfTests numeric
@ATTRIBUTE timeSpentOnLine numeric
@ATTRIBUTE avgGradeInTesting  {low, average, high}
@ATTRIBUTE clasa {weak, average, good, excelent}

@DATA


32,6,10,average,excelent
17,10,8,low,weak
20,5,4,low,average
11,8,15,low,weak
42,20,18,high,excelent
20,5,14,low,average
```

Fig. 6 Dataset snippet

```
Correctly Classified Instances      44              78.5714 %
Incorrectly Classified Instances    12              21.4286 %
Kappa statistic                      0.7102
Mean absolute error                  0.0923
Root mean squared error              0.2626
Relative absolute error             24.8084 %
Root relative squared error         60.8057 %
Coverage of cases (0.95 level)     100       %
Mean rel. region size (0.95 level)  33.9286 %
Total Number of Instances           56
```

Fig. 7 Results on the first test

```
Correctly Classified Instances        55              84.6154 %
Incorrectly Classified Instances      10              15.3846 %
Kappa statistic                        0.7925
Mean absolute error                    0.0793
Root mean squared error                0.2354
Relative absolute error               21.3169 %
Root relative squared error           54.535  %
Coverage of cases (0.95 level)       100      %
Mean rel. region size (0.95 level)    36.9231 %
Total Number of Instances             65
```

Fig. 8 Results on the second test

results obtained after performing the tests. The second test was made using a more significant number of instances than the first one.

In Figs. 7 and 8, we can analyze how the percentage of correctly classified instances increases along with the total number of instances. Results obtained from the student's feedback will permanently increase the number of instances in the training set file, and the algorithm will be able to give better results.

Let us consider the scenario in which a student enrolled in Computer Science has sent 17 messages, has taken 20 tests, spent about 30 h online platform and obtained a "high" average test. This student's instance looks like this.

In Fig. 9, the question mark denotes that the student has not already taken the exam, so this attribute is, at this moment, empty. We cannot use this instance to train the classifier, but the variable's value can be predicted in other cases. Predicting a result and then comparing it to the real result obtained by the student can give us a good measure of the algorithm's accuracy. If the student sends a message to the professor and answers back, the system will index the messages in the following way. The student's class is to be determined, and the message is linked with the obtained class, which is, in fact, a leaf in the decision tree.

Let us suppose another student is interested in reading message recommendations from the system. The student is classified the messages linked to his class are given as recommendations. The student is asked to mark as exciting or not interesting the messages he is reading. According to his marking, the message's weight is increased or decreased. The messages rating mechanism will ensure in time that highly relevant messages are provided as feedback to students.

```
                          17,20,30,high,?
```

Fig. 9 An instance representing a student

5 Conclusions

This chapter introduces how built data analysis workflows integrate machine learning algorithms intending to predict and rank the difficulty of a course and send personalized messages to students.

The proposed method for ranking the courses in terms of their difficulty considers the number of concepts that appear as new in a course. This approach considers that courses come in a linear workflow to the student, while each course has assigned a specific number of concepts. Therefore, all addressed concepts are new at the first course and starting from the 2nd course, only unaddressed concepts in previous concepts are new. The proposed mathematical model aims to obtain a ranking in difficulty and, therefore, a way to predict performance and fine-tune the difficulty of future tests.

Further, we use a decision tree model to build a message recommender system. The critical aspect is that students are labelled according to their activity, and therefore messages and topics discussed by a specific class of students may be more relevant for similar students. The prototype implementation uses decision tree classification. The accuracy of the refined classifier is 84.61%; we conclude that performed modelling may be successfully used in production.

As future works, there may be several lines of improvement. One consists of a more refined data analysis pipeline preprocessing, and hyperparameter tuning should get more attention. The second line of improvements regards strictly the algorithms that are used within the data analysis pipeline. Advanced classification algorithms such as random forest, SVM, and XGBoost of neural networks may improve accuracy. The current limitation consists of the small number of instances (i.e., students) and only solving this issue may enable further usage of the deep learning techniques. Finally, the validation methodology may also have improvements. Besides the currently accuracy-based validation, a relevance feedback evaluation would be advisable. Finally, since we deal specifically with recommended messages, using various text embeddings such as classical TF–IDF of state-of-the-art deep-learning-based methods such as BERT may open the way towards more refined versions of the message recommender systems.

References

1. Nakamura, K.K.: Estimating learners' subjective impressions of the difficulty of course materials in e-learning environments. In: APRU 9th Distance Learning and the Internet Conference, pp. 199–206 (2008)
2. Liu, J.S.: Ranking the difficulty level of the knowledge units based on learning dependency. Int. J. Dist. Edu. Technol. (IJDET), 31–43 (2012)
3. Romero, C., Ventura, S.: Educational data mining: a survey from 1995 to 2005. Expert Syst. Appl. 33(1), 135–146 (2007)
4. Cristóbal Romero, M.-I.L.-M.: Predicting students' final performance from participation in online discussion forums. Comput. Educ. 68, 458–472 (2013)

5. Moss, J., Hendry, G.: Use of electronic surveys in course evaluation. Br. J. Edu. Technol. 583–592 (2002)
6. Pamela, H., A.A.: A review of the research on student evaluation and a report on the effect of different sets of instructions on student course and instructor evaluation. Instr. Sci. **9**(1), 67–84 (1980)
7. Ozkan, S.: Multi-dimensional students' evaluation of e-learning systems in the higher education context: an empirical investigation. Comput. Educ. **53**(4), 1285–1296 (2009)
8. Sun, P.C., Tsai, R.J.: Yeh. What drives a successful e-learning? an empirical investigation of the critical factors influencing learner satisfaction. Comput. Educ. **50**(4), 1183–1202 (2008)
9. Abel, F.B.: Recommendations in online discussion forums for e-learning systems. IEEE Trans. Learn. Technol. **3**(2), 165–176 (2010)
10. Nguyen Thai-Nghe, L. D.-G.-T.: Recommender system for predicting student performance. Procedia Comput. Sci. **1**(2), 2811–2819 (2010)
11. Davies, J.A.: Performance in e-learning: online participation and student grades. Br. J. Edu. Technol. **36**, 657–663 (2005)
12. Wang, M.: Correlational analysis of student visibility and performance in online learning. J. Asynchr. Learn. Net. 71–82 (2004)
13. Garrison, D.R., Archer, T.: Critical inquiry in a text-based environment: computer conferencing in higher education. Internet High. Educ. **2**(2–3), 87–104 (2000)
14. Wang, M.J., Kang, M.: Cybergogy for engaged learning: a framework for creating learner engagement through information and communication technology. In: Engaged Learning with Emerging Technologies, pp. 225–253. Springer, Dordrecht (2006)
15. Purdilă, V.: Fast decision tree algorithm. Adv. Electr. Comput. Eng. **14**(1), 65–68 (2014)
16. Asif, R.M.: Analyzing undergraduate students' performance using educational data mining. Comput. Educ. (2017)
17. Daud, A.A.: Predicting student performance using advanced learning analytics. In: Proceedings of the 26th International Conference on World Wide Web Companion, pp. 415–421
18. Iwata, T., Tanaka, T.: Improving classifier performance using data with different taxonomies. Know. Data Eng. IEEE Trans. **23**(11) (2011)
19. Fuxman, A., Kannan, A.: Improving classification accuracy using automatically extracted training data. In: KDD '09 Proceedings of the 15th ACM SIGKDD International Conference on Knowledge Discovery and Data Mining. ACM (2009)
20. Joshi, M.T., Agarwal, R.C.: Evaluating boosting algorithms to classify rare classes: comparison and improvements. In: Proceedings IEEE International Conference Data Mining, ICDM 2001. IEEE (2001)
21. Bryll, R., Ricardo, G.-O.: Attribute bagging: improving accuracy of classifier ensembles by using random feature subsets. Pattern Recognit. **36**(6), 1291–1302 (2003)
22. Minaei-Bidgoli, B., Kashy D.: Predicting student performance: an application of data mining methods with an educational Web-based system. In: 33rd Annual Frontiers in Education, FIE 2003, vol. 1, pp. T2A-13. IEEE (2003)
23. Popescu, P.Ş., Mihăescu, M.C.: Automatic course difficulty evaluator for Tesys e-learning platform. In: Proceedings of the18th International Multiconference Information Society is 2015, vol F. Ljubljana (2015)
24. Burdescu, D.D., Mihaescu, M.C.: TESYS: e-learning application built on a web platform. In: ICE-B, pp. 315–318 (2006)
25. Quinlan, J.R.: Decision trees and decision-making. IEEE Trans. Syst. Man Cybern. **20**(2), 339–346 (1990)
26. Hall, M., Frank, E., Holmes, G., Pfahringer, B., Reutemann, P., Witten, I.H.: The WEKA data mining software: an update. ACM SIGKDD Explor. Newsl. **11**(1), 10–18 (2009)
27. Mocanu, M., Popescu, P.Ş., Burdescu, D.D., Mihăescu, M.C.: Advanced messaging system for online educational environments. In: Intelligent Interactive Multimedia Systems and Services, pp. 61–69. IOS Press (2013)
28. Popescu, P.Ş., Mocanu, M., Burdescu, D.D., Mihăescu, M.C.: Messaging activity impact on learner's profiling. In: 2015 6th International Conference on Information, Intelligence, Systems and Applications (IISA), pp. 1–6. IEEE (2015)

Building Interpretable Machine Learning Models with Decision Trees

P. S. Popescu⑩**, M. C. Mihăescu, and M. L. Mocanu**

Abstract Interpretable machine learning models is a subject that remains relevant for educational data mining even if it started once a long time ago because when working with students in online educational environments, we need to understand them better and to lower the gap between full-time education and distance learning. In order to improve the interpretability of the models trained on educational data, we propose two approaches that can be used for model analysis and to provide recommendations. Both approaches are based on decision trees, and even if the experiments are focused on the Tesys e-Learning platform, they can be replicated in almost any other context. The first approach is based on leaf analysis and automatically tries to find and recommend tutors for low learning level students. The second one refers to students' ranking based on leaf analysis; this ranking considers all the attributes that describe a student as an instance from the dataset. The experiments from the final part of this chapter validate the systems and provide more details for a deeper understanding.

Keywords Educational data mining · Machine learning · Decision trees

1 Introduction

Using machine learning algorithms to improve the interaction and learning experience in both online educational environments and full-time education is a technique that has become popular. One specific problem is that not all the machine learning algorithms offer interpretable models, most of them providing just a result, usually for a classification problem. In this case, the standard evaluation metrics play a significant role in evaluating the system's performance, but it is hard to evaluate how generic the system is and know how it will perform on unseen data. The data used for educational data mining also has specific characteristics, usually having a small number of instances, attributes that may not be relevant for a specific task or

P. S. Popescu (✉) · M. C. Mihăescu · M. L. Mocanu
Department of Computer Science and Information Technology, University of Craiova, Str. A.I. Cuza, Nr. 13, Craiova, Romania
e-mail: stefan.popescu@edu.ucv.ro

unbalanced classes. The attributes and their importance in the final machine learning model usually influence the model bias and how it performs because if the machine learning model is based on relevant attributes (e.g., in a decision tree algorithm we have as root and on the upper levels relevant attributes) then the results are also relevant in that context. The opposite situation is that we have a model that performs exceptionally well and offers good performance on the training dataset, but on the unseen data, we may not get the same accuracy because the attributes may not be relevant for the class.

Interpretable machine learning models are essential for preventing overfitting and further analysis, which can reveal necessary information about the instances and, more precisely, the students who use the online learning environments. A good scenario is to build a dataset based on the data gathered from an e-Learning or learning resource management platform and train a basic decision tree algorithm. In this scenario, if we have a small number of features in our dataset, we can visualize the tree and see which attributes provide the best information gain for what values the splits are done. Starting from the root, parsing the tree enables us to evaluate the model from a data analyst perspective, and then we can say if the tree is relevant for our desired classes, or we should tune the dataset or the algorithm, but this approach may not work for each case. Nevertheless, there are several cases in which such an analysis may not be feasible because of the dataset, which can have too many attributes, or because of the model, which may be too large and complex to be parsed visually. Too many leaves in which there are few instances even if the dataset has a small number of features and instances is a bad situation. However, these are the cases in which the algorithm can provide a visual model; there are also cases in which we cannot construct such a model because of the way the algorithm works.

Having interpretable machine learning models lets us implement other systems in our e-Learning platform—Tesys, like a tutor recommender system and a student's ranker, presented in detail in this paper. These two systems aim to enhance the user's experience in the e-Learning systems and improve the interaction between users, which is a big difference between full-time education and online learning environments. The importance of interaction between users is fundamental as it can improve knowledge transfer and engagement in a learning activity, leading to better learning performance.

The tutor recommender system starts from the idea that students have different levels of knowledge, and their learning performance may improve if we find a way to improve their interaction through the e-Learning platform. Our system starts from a classification problem in which we classify students in three classes: low, average, and high and for a model building, we train a decision tree classifier. This classifier is used not only for classification purposes but also for leaf analysis, and we can recommend tutors from leaves that have many things in common with students classified as low by our trained model. We consider that matching low-level students with high-level students who share many learning characteristics can improve learning performance, and the high classified students can provide an excellent tutoring experience for the low classified ones.

The other system presented in detail in this chapter refers to student ranking using a classification algorithm and aims to trigger an early alert if a student is in danger and could fail at the final exam. The motivation for using a classification algorithm for ranking and preparing a dataset, especially for that, instead of ranking them using the grades obtained until that checkpoint, is that learning performance and knowledge level are not defined only by grades. Collecting more data about a student can reveal his true strengths, weaknesses, and can provide a clearer image of his actual learning status. In this case, we also used a decision tree algorithm, and we trained it on an especially prepared dataset which follows the rule that each attribute follows a rule that if its values have an ascending order, their significance goes from low to high. For example, if we have an attribute x with values ranging from 1 to 10, 1 will be the lowest/worst value, and ten represents the best. Using such attributes on our specifically chosen decision tree model will produce an output in which each leaf is ranked from left to right, going from low to high. The outcome of the system is that we can parse the leaves and see the students from them, see their place in the ranking, and read the rule that led to constructing that leaf. This rule is essential to be composed by the attributes and their split, and we can see which attributes a student is not performing well enough to give him recommendations for improving his learning performance.

For model interpretability improvement and in order to be able to build the systems presented in this chapter, we also implemented several functionalities like leaf parsing, computing successors and predecessors of a leaf, assigning instances to an already built model and others. All these functionalities, along with the above-presented systems, will be presented further in this chapter.

2   Related Work

2.1   *Background Related to View Techniques for Better Model Analysis*

Building models that provide researchers with excellent and reliable feedback regarding the data is a must in intelligent data analysis. In some cases, understanding the model can bring additional information to the data analyst and also, some patterns can be observed. These patterns are not explicitly provided by the algorithm, or the model, as it was not designed for that, but the data analyst can see them and use them further. Understanding the model is beneficial to understand the results and the pre-processing step in which some features can be eliminated or combined for better model accuracy.

One paper published by [1] presents LOTUS, a comprehensible algorithm based on logistic regression trees. The paper proposes an approach that recursively partitions the data and fitting logistic regression in each partition, so the results will be a binary tree that is pruned after that. The authors assume that the model is easy to

interpret if a simple linear logistic regression is settled on each partition. The experimental results presented in the paper is pretty huge; 30 real datasets were used, and
they tested this method against standard logic regression and found that this method
reduced the predicted mean deviance by 9–16%. They also took an example for the
Dutch insurance industry and demonstrated that the method could produce an intel-
ligible profile of prospective customers. As it may seem that their approach is quite
common, and the algorithm is similar to M5 [2] or CART-Logit [3], but in this case,
both of them require building a tree as a preliminary step, and the usual exhaustive
search is very exhaustive for logistic models. In our approach, which is included in
this chapter, the search is performed only at the leaf level, and it is not only faster
because it does not take into consideration all the nodes, but also the presentation of
the results is more interpretable than the search included in [4].

2.2 Related Work for Innovative Ways to Rank Instances

Even very recent research [5] that analyses learners' performance using a data
mining approach uses classification (decision tree) models or clustering to predict
the learners' results. This approach is fundamental because a significant amount of
information can be lost using only class analysis. The authors state that using decision trees combined with the Gini index can have a better prediction accuracy, but
in most cases, the accuracy obtained after cross-validation is the only measure that
can give us an overview of how well a classifier performs. The results presented in
the paper refer only to the classification accuracy, and even so, the average value is
around 70% which is not very good.

ML models and their interpretation have always been an important task. One paper
[6] aims to predict students' performance using advanced learning analytics, and the
authors also present visualization techniques. The experimental results are done using
Weka 3.7, but the dataset has only 100 instances and 23 features. The feature ratio
regarding the instances is undoubtedly not the best. The performance evaluation is
performed using precision, recall, and F1 score despite computing accuracy like in
most cases. One drawback is that the construction of the dataset might be biased
because of the selection of the students (50% completed, 50% dropped), and also
their approach using above mentioned precision metrics does not offer noticeable
results.

In this research area, we can improve the algorithm's performance by adding more
functionalities because this approach can increase the algorithm's flexibility.

An interesting approach for improving the classifier's performance is presented in
the paper "Improving Classifier Performance using Data with Different Taxonomies"
[7]. In [7] the authors propose a framework for increasing the classifier's performance by efficiently using supplementary samples. The extra samples are labelled not
regarding the target taxonomy according to which they want to classify instances but
according to classification methods or taxonomies different from the target taxonomy.

Their approach suggests a classifier by lowering a weighted error over the target and auxiliary instances.

Another method for enhancing the classifier's accuracy [8] is features bagging or boosting techniques as presented in [9]. In paper [10], the authors present a method of using the features bagging technique in order to improve both the accuracy and the consistency of classifier ensembles. This approach is induced by using random subsets of features. Features bagging is a generic approach which can be used in any machine learning algorithm. Their approach establishes a custom subset of features, and then it arbitrarily selects several subsets of attributes, creating projections of the training data set on which the ensemble classifiers are trained. The trained classifiers are then used for voting technique. That article compares the performance of their approach with bagging and other algorithms on a hand-pose recognition dataset. It is presented that their approach offers consistently improved outcomes than bagging, both instability and accuracy.

Plenty efforts [11] have been made in the Educational Data Mining research area only for integrating data mining and machine learning algorithms [12] in the educational research area [13]. In this chapter, we choose to present the analysis process and several features implemented for decision tree data mining algorithms to improve their performance and extend their usage area. This chapter is strongly related to educational data mining research areas as we developed an advanced classifier that was validated on educational data and aims to improve students' learning experience. Regarding this, there was an earlier paper [14] that presents an exciting approach that aims to classify students to predict their final grades. The features were extracted from the data logged in an education Web-based system. The authors design, implement, and evaluate many pattern classifiers and perform a benchmark comparison on their performance on a dataset obtained from an online course.

2.3   Related Work on Building Interpretable Models

Newer papers also analyze interpretable models applied in e-Learning; a survey being published in this direction in 2019 [15] and newer papers like [16] present interpretable models in the COVID-19 Pandemic context. The research presented in [16] contributes to the literature of online collaborative learning between educators, parents, and schools that impact learners' success. The result of this analysis presents four main themes, instructional strategies, challenges, support, and motivation of lectures. E-learning platforms were used as solutions to refer mainly to complete online education between full-time education and online educational environments learning outside the campus facing covid-19. Another new paper [17] that focuses on the same problem uses advanced machine learning models such as deep learning networks and gradient boosted machines. They aim to predict the final academic results of students at the Budapest University of Technology and Economics. The failure prediction is based on the data available when students are enrolled. In addition to the forecasts, they also explain their machine learning models with the help of the most used interpretable machine learning techniques like permutation importance and SHAP values.

Regarding the results, their accuracy and AUC of the best-performing deep neural network model is 72.4% and 0.771, and that slightly outperforms XGBoost [18] algorithms which is the best and most used benchmark model for tabular data. Another recent paper [19] which tackles the same problem of interpretable models [20], explores the possibility of using deep learning models in Learning Management Systems [21]. Their focus was on discovering knowledge about the learning process using dataset analysis performed by Artificial Neural Networks (ANNs). They conclude that although ANN models provide high accuracy, the result is an uninterpretable black-box model, which is a significant drawback. In order to better understand the ANN black-box model, as well as for other models of this type, several specific methods have appeared recently, some of which they illustrate in their LMS system analysis.

2.4 Weka

For most of the experiments performed further, we use Weka, an open-source library of machine learning data mining and algorithms developed in Java. The included algorithms are pretty flexible and maybe run in the analysis process of different types of data (and from any domain). Weka has three main types of algorithms: unsupervised or grouping algorithms (e.g., partitioned, EM [22], fuzzy [23]), supervised (e.g., decision trees, Bayesian networks [24], vector space classification [25]), and association rules like the "Apriori" algorithm.

The motivation for using Weka despite other more common packages is that we implement our systems for Tesys e-Learning platform, which is also developed in Java and using such a library simplifies the process a lot. Another reason is that we can use the standard algorithms with a standard set of parameters, but we can also set specific parameters for better performance to be used from a straightforward approach to an as low level as possible.

Weka offers a tool with an interface allowing us to do experiments without development and a. jar library that can be included in Java projects, so it is easy to integrate and develop custom systems using it. Because the library allows contribution and code extension, there is a core app that comes with the most used algorithms and a list of available official packages which anyone can install or integrate into his or her code. There is also de possibility to build packages and contribute to the community.

3   Design of the Proposed Techniques

3.1   *Design of a View Technique for Better Model Analysis*

The paper that supports the design of this view technique is [26] and does not represent a direct way to produce recommendations for students or professors but having such a data structure allows us to make a subtler refinement at leaf level to produce more accurate recommendations. Another benefit of building and analyzing such a data structure is that depending on the features used, we can have an ascending order of leaves, so we can implement functions that allow a data analyst to parse the tree and get a better intuition regarding the instances based on the leaf position in which they belong.

Figure 1 describes the core functionality of the system and how modules interact at a high-level design. In this figure, there are several levels in this image, each meaning a step in the development part.

On the first level of the design document, we have the data set and the Classification algorithm. This part is done using the Weka library; the data set is represented by a training file that can be parsed by Weka's algorithm, which in our case is a ".arff" file. Using the classification algorithm, we can go to the next level to compute the model that is not loaded with data. This part could be done without too much effort.

We will have our desired Dense decision tree by taking a current dataset, which can be even the training dataset, and applying the data loading algorithm that loads the model with data. On this model, loaded with data, we can compute several techniques that will allow us to see the predecessor or successor of a specific node and also, we will be able to compute the ith successor or predecessor of a specific leaf.

For predecessor computing, we have a *computePredecessor* function similar to the one mentioned above and is presented in Sect. 4 of this chapter. The parameter significance is the same, and if we want to refer to the differences, we can find

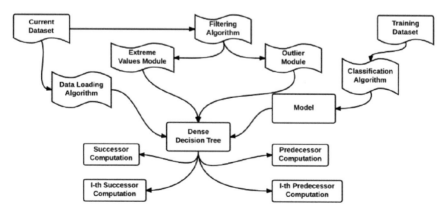

Fig. 1 High-level system design

them in the node's iterations and an extra check to see if we iterated at least the predecessor's rank when we parse the list of leaves. On the other hand, applying a filtering algorithm enables us to find the outlier and extreme values found within the current dataset.

Later, based on the previously presented approach, a paper entitled "J48 list ranked based on advanced classifier decision tree induction" was published, and it represents the next contribution for this research goal. As mentioned before, building an advanced classifier that encapsulated several functionalities enables us for further and more accurate data analysis. Regarding model analysis, having a preloaded tree and the possibility of getting the rules that assigned an instance to a leaf can provide a better intuition regarding that instance because, in most cases, more leaves belong to the same class. However, there are different rules, and these rules analyses can reveal valuable information.

Improving classifiers is done by taking one of two main directions: redesigning the algorithm for better performance or adding more functionality that extends the range of applications and makes the algorithm more flexible. One of the examples of application domains in which we consider to be useful is an educational research area. Based on this context, students are defined by their actions made in e-Learning platforms. These actions are logged mainly in "numeric" data type, which is ordered. Based on information gain computed at every step by the machine learning algorithm, we can see that the bigger values for the split nodes are on the right side of the tree, and the lower values are on the left. From this point of view, the smallest values indicate the worst position in the ranking, and large values indicate the best position.

Thinking that every attribute is related to the others by the meaning of the values, we can gain more particular importance for the leaves. Traversing the leaf-list in one approach which can bring us more exciting information about the student ranking and recommend him tutors or learners with better but related characteristics.

This idea of ranking instances based on leaves is presented in Fig. 2. In the first row, we have the values of the features ranked in numerical order, and in the second row, we illustrate the relation between values and ranking. In this Fig. 2, we can observe that instances (from the data file) that have lower values for their features

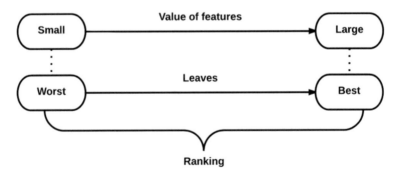

Fig. 2 Relation between value and ranking

belong to the leaves that correspond to the "worst" value in this ranking scenario and instances that have bigger (more significant) values to the "best" (in the right side of the tree. This statement occurs because the J48 classification algorithm will perform in a specific manner on well-formatted datasets. We can use this machine learning algorithm as a ranker because of the algorithm for data loading that allows us to load leaves with instances.

Our developed ranking algorithm is quite generic, but the dataset needs to comply with some limitations. These limitations are in a close relationship with the obtained ranking model. This ranking is suitable for data types like numeric and nominal. For numeric features, the limitation is that lower integer, or real values, are placed on the left side of the split node, and bigger values are placed on the right side of the node in which the split is made. Based on the same approach, the first nominal value is placed on the left side of a node, and the last nominal value is placed on the right side of a node. Regarding the values significance, values from the left side of the node are associated with lower educational performances, and the same rules also applies for class attribute, but we have to specify that meaning of low or bad and high or good is relevant on the context from which the data was extracted.

One last limitation comes from the datasets is that we cannot load the tree with a dataset that has a different structure from the one that was used for training. This dataset must need to have the same number, name, and type of the features used in the training dataset; we may have missing values, but the attribute needs to be there.

In order to work this approach, there are several problems that need to be solved. One problem was to develop a generic tool for attributes parsing which is not dependent on the data types. Our approach to this difficulty was to save the obtained decision tree model into a set of rules, and each rule being constructed as the path from root to each leaf. This permits us to be able to fill the dataset in a list of hash maps, each component of the list containing the instances that belong to that leaf. From the programming perspective, the best option was to split the objects that define a tree into different data types and use them further in the tree for data saving. We create in our tool three different classes: one for the internal nodes, one for the leaves, and one for the root, each of these classes having a list of objects from the lower rank (i.e., root has a list of internal nodes).

At the leaves level which represents the bottom of the tree, we have the most important component from our structure because we save a list of instances which can be in the leaf and a hash-map that saves the rule computed parsing the tree from the root to that leaf. For each leaf of the tree model, we compute a composed rule, and we save it as a range of values for each attribute from the dataset. When we want to load an instance to a leaf, a set of rules will be verified so the instance will be placed in the right place.

Another challenge we faced was to rank the obtained list of leaves. Taking into consideration that the features need to comply with the structural limitations imposed by the model we can perform a depth-first search in our classification model in order to obtain the correct order when the leaves are loaded into our ranked list model. When we traverse the tree, we visit the children nodes from small to large feature values in order to follow the structural data constraints imposed by the algorithm.

Pseudocode:

Procedure: buildRanking(data model)

    While (ordered DFS traversal of decision tree)

        # insert leaves to ranked list

    For each (item in test dataset)

        #add item to corresponding element in ranked list

    Return ranked list

Fig. 3 Overview on the algorithm for building ranking

In Fig. 3, we present the pseudocode for the algorithm used for loading the data into the ranked list. This approach allows us to create a generic development that works nicely on our test scenario but also in any other scenario. In this case for each instance from a dataset we want to load we have two steps: first we test it using the computed rules until we find the leaf that accommodates it, then we add it to the list of ranked instances. For this instance, ranking algorithm based on decision trees model, we have built a data processing pipeline which takes as input the data model along with a set of instances and outputs a ranked list of instances based on the decision tree model. In order to make the system flexible and open for further development, the data loading algorithm was designed to work with decision tree algorithms that have their models saved (or converted) in XML format. The whole dataflow used for ranking is presented in Fig. 4.

Figure 4 presents the data processing pipeline, and we have as starting point the algorithm from Weka's library which is a basic decision tree algorithm (i.e., j48) that can output us a decision tree in *.txt* format. Generically when we use this text decision tree model, we can easily convert it further in an XML decision tree that can be more simply used to compute the rules and after the rules are computed it will be easily used to be loaded with instances. After using our algorithm for loading data, we can obtain a new XML file that has the same structure with the decision tree model, but this time is loaded with the instances from the dataset. We can analyse this model with instances in order to better understand our students, or we can apply it to several other operations like computing the predecessor or successor of a desired leaf or computing different performance and validation metrics which can provide us an overview of the how good the algorithm works.

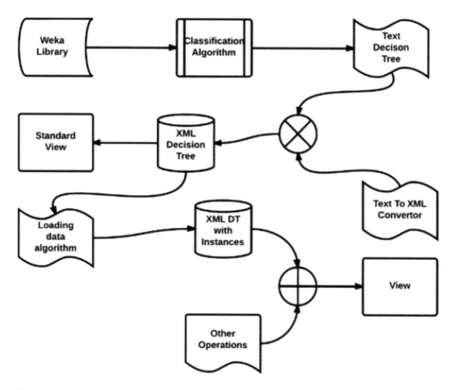

Fig. 4 Data flow

4 Experiments and Results

4.1 Results on the View Technique for Better Model Analysis

The paper that presents the experiments and the design and conclusions in details for these experimental results is [26] and should be consulted for further explanations. A standard decision tree algorithm will be able to provide only the class that an instance belongs to but taking into consideration two classes, "A" and "B", we can have several instances that are assigned to class "A" but are different in several ways between them (being in different leaves). For instance, we can have some students classified as "A," some of them perform well at physics, and some of them perform well at math. Based on the trained decision tree, both students are correctly classified but let's consider the problem that we want to find the best students on math efficiently. We may use a different training dataset, train another algorithm that may perform better for that particular case or we can have the same one doing this job and we can have even more functionalities.

Considering a more specific way, we can find tutors or students that can help other students to improve their learning performance. This can be accomplished by parsing the tree and then finding successors of a specific leaf. There are two situations in this case: the next successor (leaf) is the desired class (high class), and finding is over and one situation in which we have to search further in the leaves list.

Taking into consideration a set of features that can define a particular student and parsing the resulted tree, we will always have a sibling or a successor for the leaf in which is the student, and that sibling will be in high class. If we have a sibling, we will can a tutor who is very similar and close to the student that needs help. In case we do not have a sibling for that specific leaf, we can traverse further for successors, and we can also rank the level of similarity of the next student. In this case, the matching mechanism between students is more exciting because in most of the cases there is not only one attribute that is different between those classes. If we just choose the next successor, it is most likely to find a leaf with students that are also classified as low, but they performed better at the level that we had the split in the model. If we consider that we want only students that perform well to become tutors, we will need to iterate at least twice processing the successor of the leaf that has the student with difficulties.

The ith successor or predecessor opens some new research perspectives. Regarding our educational area in which we aim to match students, we have one specific problem that will be discussed in what follows. Let us consider that in Fig. 5, we have students which were loaded in leaf K, and we want to search some students that may help them. If we looked only for successors, we would get the students from leaf L, which can be the only average if we consider three classes of students: low, average and high. Choosing leaf L as source for tutors may lead to students that are closer to class low (we consider that leaf K is in class low), these students certainly

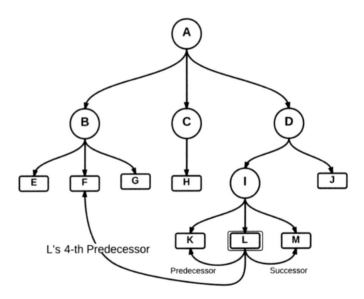

Fig. 5 Example of successor and predecessor

have more in common with those from leaf K, but there may be some situation in which they will not help. Computing the next successor will provide us the students from leaf M, which is rated as high; these students are the most likely relevant for the students from leaf L being close to them but in a high class and being the best from that situation.

In another specific case is that in which there can be a situation when we want to find tutors for students that are in a leaf ranked as high. This is a bit more difficult and needs more recursive processing to parse the tree and find where is done a split at a higher level and then search for the leaf which is ranked as high. We make this selection because the students from the actual class are also ranked as high, but somewhere at a higher level in the tree, there was a decision where he went on the left side, which most of the time means those students had a lower ranking. If we use the J48 machine learning algorithm, the high rightmost leaf will be the only one cannot get a recommendation with better students.

Regarding outlier and extreme values computation, if we focus on the educational data, we can find two types of students that would be very interesting to analyse. These students are very distinct from the other students, and they can be excellent students or, in some cases, deficient students. Both of these types of students are important because they need to receive special care and even more attention. The students ranked as excellent can frequently be introduced in special activities that will allow them to better participate in special programs, to support them and find ways to motivate them or be guided in order to achieve a successful learning direction.

4.2   Short Dataset Example

For example, we can consider the following sample of dataset:

@ relation StudentClass.
@ *attribute nrHours numeric.*
@ *attribute avgMark numeric.*
@ *attribute messagingActivity {YES, NO}.*
@ *attribute noOfTest numeric.*
@ attribute messageLenght {short, long}.
@ *attribute class {low, high}.*
@ *data.*
51, 4.5, YES, 4, long, high.
29, 6.025, NO, 4, long, low.
14, 5.1666665, YES, 6, long, low.
30, 3.25, NO, 5, long, low.
18, 6.75, YES, 2, short, low.
51, 6.4166665, NO, 6, long, high.

After applying the InterquartileRange algorithm (filter in weka), we will have two more features added to our dataset; with these features, we can indicate if an instance can be considered an outlier or having an extreme value. For the dataset presented above, we will get the following end result:

@relation StudentClass.
@attribute nrHours numeric.
@attribute avgMark numeric.
@attribute messagingActivity {YES, NO}.
@attribute noOfTest numeric.
@attribute messageLenght {short, long}.
@attribute class {low, high}.
@attribute outlier {no, yes}.
@attribute ExtremeValue {no, yes}.
@data.
51, 4.5, YES, 4, long, high, no, no.
29, 6.025, NO, 4, long, low, no, no.
14, 5.166667, YES, 6, long, low, no, no.
30, 3.25, NO, 5, long, low, no, no.
18, 6.75, YES, 2, short, low, no, no.
51, 6.416667, NO, 6, long, high, no, no.

We used data from the educational research area for our test case scenario, and in this case, we could not find any extremely high or deficient students. This case can be rare for our context, but considering that our tool can be used widely in different domains, we can get this data as a data analyst computing these variables before model loading can positively impact it. For example, in the tutor's usecase, we want to remove the highly gifted students from the decision tree model in order not to harm them with the tutoring activity.

Finally, for visualization our loaded decision tree model, we employed a JTree loading it with instances and marking an actual class of a particular student with red and the needed class (the class of tutors, or target class) with green as shown in Fig. 6.

After each leaf, which has the identifier "Class" in front of it, we have line called "Students," here are the IDs of the students that belong to that specific leaf. The IDs of the students correspond to the ones from the database.

4.3 Validation of the Procedure of Ranking Instances Based on Leaf Analysis

After developing an advanced dense classifier, in paper "J48 list ranker based on advanced classifier decision tree induction", we found that the previously presented advanced classifier can rank the classified instances for specifically formatted datasets. For our experimental results, we utilized a dataset that has 787 instances

Fig. 6 An output examples

```
Applet
* DecisionTree J48Tree
? nrHours <= 35
  ? avgMark <= 6.66
    ? Class: low -> 340 students
        Students:
  ? avgMark > 6.66
    ? avgMark <= 7
      ? Class: low -> 13 students
          Students: 1 8
    ? avgMark > 7
      ? Class: high -> 10 students
          Students: 12
? nrHours > 35
  ? messageLenght = short
    ? messagingActivity = YES
      ? Class: high -> 40 students
          Students: 2 6
    ? messagingActivity = NO
      ? Class: low -> 44 students
          Students: 9
  ? messageLenght = long
    ? avgMark <= 4.25
      ? avgMark <= 4
        ? Class: low -> 5 students
            Students: 10
      ? avgMark > 4
        ? Class: high -> 5 students
            Students: 11
    ? avgMark > 4.25
      ? avgMark <= 7
        ? Class: high -> 165 students
            Students: 3 7
      ? avgMark > 7
        ? noOfTest <= 3
          ? Class: low -> 3 students
              Students:
        ? noOfTest > 3
          ? Class: high -> 6 students
              Students: 4 5
```

defined by four features. This dataset uses real data collected from the students performing their learning activities on the Tesys e-Learning platform.

For this experiment, we considered using the information regarding the last three years of study from the database. This method provides a good environment in which we can evaluate and validate the whole system. For this method, we have chosen three attributes that were computed from the database. The class attribute was computed regarding the student's result because when we create the decision tree model, we refer to students that have already graduated studies.

In Fig. 7, we can see a sample from the *arff* file with attributes and some instances used for defining a student. We need to mention that attributes are proposed only for testing purposes and are not the only attributes which can be gathered for describing a student.

In our experiment:

- *"nrHours"*—represents the total number of hours spent by a student on the platform doing learning associated actions. When we say learning, we refer using

Fig. 7 Sample snipping of
the training dataset

```
@RELATION StudentClass

@ATTRIBUTE userid numeric
@ATTRIBUTE nrHours numeric
@ATTRIBUTE avgMark numeric
@ATTRIBUTE present {NO,YES}
@ATTRIBUTE class {low,high}

@DATA
4,100,10,NO,low
5,11,7.0,YES,low
7,30,7.375,NO,low
8,10,7.714286,NO,low
9,500,1,YES,high
10,31,4.5,YES,high
11,11,3.25,NO,low
12,44,6.4166665,NO,low
```

learning resources like reading courses, doing homework, and taking tests for
training. The range of the values goes from 0 to 100, and we consider 0 value to
be low and 100 to be the highest value.

- "*avgMark*"—refers to the marks obtained at tests during the semester of study.
  The range of the values goes from 1 to 10, 1 meaning the lowest mark and 10 is
  the best.
- "*present*"—feature describe the data for attending the online presentations. We
  consider it to be "NO" if the student attended few or none of the presentations
  (lowest value) and "Yes" if he attended most of the presentation (highest value).

The tree model presented in Fig. 8 is the result obtained by training the algorithm
with the above dataset. We can observe here the number of instances assigned to
each leaf after loading them and the order of the items related to the leaf's succession. Traversing from left to right, we obtain a list of items from the lowest to the
highest level of performance obtained by the instances in them. For this approach,
the instances are represented by students, and we aim to rank them by academic
performance. In order to validate the model, we load the tree model with a set of 25
students (who graduated) and analyse how the algorithm works with this data.

In Table 1, we have the leaves loaded with students, on the second row we have
the final grade obtained by each student id and on the last one the average mark of the
students from that leaf. Examining this table, we can see that looking at the leaves
Leaf 1 and *Leaf 2*, only one student had a final grade above the students from *Leaf 2*.

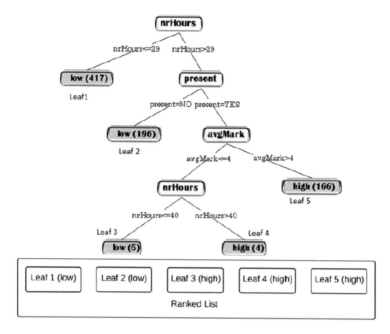

Fig. 8 Sample ranked list using a decision tree

Table 1 Students loaded in leaves

Data	Leaf				
	Leaf 1	Leaf 2	Leaf 3	Leaf 4	Leaf 5
	Low	Low	Low	High	High
Student ID	1, 4, 10, 21, 24	3, 8, 14, 19, 22, 23	7, 9	11, 12, 17, 30	2, 5, 6, 13, 15, 16, 18
Final grade	3, 4, 6, 2, 4	3, 6, 2, 4, 4, 5	6, 7	8, 7, 9, 9	9, 10, 8, 8, 9, 10.9
Average mark	3.8	4.63	6.5	8.25	9

When comparing *Leaf 2* and *Leaf 3*, there is a perfect ranking because all the students from *Leaf 2* had grades lower or equal to the ones from *Leaf 3*. Going further with our evaluation we compare *Leaf 3* and *Leaf 4*, and, in this case, we have two students which share the same grade, and here we can assume that our ranking algorithm found a difference between them because of the splits produced by features from at upper levels in the tree model. The variations between the last two leaves (*Leaf 4* and *Leaf 5*) are not significant enough having more students which had the same grade and most of the students from these two leaves have values between 8 and 10 as a final grade, and only one has got a 7 (from *Leaf 4*). Computing the average mark among the students from each leaf, we have an average of 9 for leaf 5 and an average of 8.24 for leaf 4 so there is an overall better mark for leaf 5 even if the difference is not very significant.

Examining the table, we can conclude that the overall mean final mark computed over the students that belong to each leaf is ascending linear once with the leaves order. Based on this algorithm, we can also compute ranking over the classes and this ranking can be based on the average mark of the final mark calculated over each leaf. In this case we can observe that students who belong to leaves classified as low-class have lower marks than the students that were assigned to the leaves classified as high-class leaves (4.97 vs 8.62). One more thought is that, students considered low may not pass the final evaluation exam, but those who received a high label most likely will.

5 Conclusions

This chapter presents an approach for building interpretable machine learning models. This approach refers mainly to educational data extracted from Tesys e-Learning platform and uses for model building decision trees that offer excellent and fast results and easy to validate models. Our approach focuses on leaf analysis and can offer a better student understanding, a ranking algorithm that can enhance the classroom overview and a tutor recommender system. Each of these systems is based on interpretable models and enhances the model's interpretability by adding functionalities that can better understand the context. One adjacent benefit of having a tutor recommender system for an e-Learning platform is that it can improve users' interaction, which makes a difference between full-time education and online educational environments. The results presented in this chapter are relevant for our context, and based on the datasets we used, these results can be pretty generic and easy to replicate in other systems because they have a short list of attributes that can be gathered from any learning environment.

As future research, there is plenty of work as ensemble models became more popular and offered great results but with models that can hardly be parsed or analyzed. We aim to extend our research on such models and find ways to make them more interpretable and easier to understand.

References

1. Loh, K.-Y.C.-Y.: LOTUS: an algorithm for building accurate and comprehensible logistic regression trees J. Comput. Graph. Stat. **13**(4), 826–852 (2004). https://doi.org/10.1198/106186004X13064
2. Quinlan, J.R.: Learning with continuous classes. In: Proceedings of AI'92 Australian National Conference on Artificial Intelligence, pp. 343–348. World Scientific, Singapore (1992)

3. Steinberg, D.A.: The hybrid CART-logit model in classification and data mining. In: Eighth Annual Advanced Research Techniques Forum. American Marketing Association, Keystone, CO (1998)
4. Mihăescu, M.C., Popescu, P.Ş., Burdescu, D.D.: J48 list ranker based on advanced classifier decision tree induction. Inter. J. Comput. Intell. Stud. **4**(3–4), 313–324 (2015)
5. Asif, R.M.: Analyzing undergraduate students' performance using educational data mining. Comput. Educ. (2017)
6. Daud, A.A.: Predicting student performance using advanced learning analytics. In: Proceedings of the 26th International Conference on World Wide Web Companion, pp. 415–421 (2017)
7. Iwata, T., Tanaka, T.: Improving classifier performance using data with different taxonomies. Knowl. Data Eng. IEEE Trans. **23**(11) (2011)
8. Fuxman, A., Kannan, A.: Improving classification accuracy using automatically extracted training data. In: KDD '09 Proceedings of the 15th ACM SIGKDD International Conference on Knowledge Discovery and Data Mining. ACM (2009)
9. Joshi, M.T., Agarwal, R.C.: Evaluating boosting algorithms to classify rare classes: comparison and improvements. In: Proceedings IEEE International Conference Data Mining. ICDM 2001. IEEE (2001)
10. Bryll, R., Ricardo, G.-O.: Attribute bagging: improving accuracy of classifier ensembles by using random feature subsets. Pattern Recognit. **36**(6), 1291–1302 (2003)
11. Romero, C., Ventura, S.: Educational data mining: a survey from 1995 to 2005. Expert Syst. Appl. **33**(1), 135–146 (2007)
12. Baradwaj, B.K., Pal, S.: Mining educational data to analyze students performance. Int. J. Adv. Comput. Sci. Appl. **2**(6) (2011)
13. Romero, C., Ventura, S.: Data mining algorithms to classify students. In: Proceedings of the 1st International Conference on Educational Data Mining (EDM'08) (2008)
14. Minaei-Bidgoli, B., Kashy, D.: Predicting student performance: an application of data mining methods with an educational Web-based system. Front. Educ. (2003)
15. Mduma, N., Kalegele, K., Machuve, D.: A survey of machine learning approaches and techniques for student dropout prediction (2019)
16. Elihami, E., David, V., Melbourne, M.: The interpretable machine learning among students and lectures during the COVID-19 pandemic. Jurnal Iqra': Kajian Ilmu Pendidikan **6**(1), 27–38 (2021)
17. Baranyi, M., Nagy, M., Molontay, R.: Interpretable deep learning for university dropout prediction. In: Proceedings of the 21st Annual Conference on Information Technology Education, pp. 13–19 (2020)
18. Mitchell, R., Adinets, A., Rao, T., Frank, E.: Xgboost: scalable GPU accelerated learning (2018). arXiv:1806.11248
19. Matetic, M.: Mining learning management system data using interpretable neural networks. In: 2019 42nd International Convention on Information and Communication Technology, Electronics and Microelectronics (MIPRO), pp. 1282–1287. IEEE (2019)
20. Murdoch, W.J., Singh, C., Kumbier, K., Abbasi-Asl, R., Yu, B.: Interpretable machine learning: definitions, methods, and applications (2019). arXiv:1901.04592
21. Paulsen, M.F.: Experiences with learning management systems in 113 European institutions. J. Educ. Technol. Soc. **6**(4), 134–148 (2003)
22. Bradley, P.S., Fayyad, U.M.: Reina scaling EM (expectation–maximization) clustering to large databases. Microsoft Res. (1999)
23. Looney, C.G.: A fuzzy clustering and fuzzy merging algorithm. Tech. Rep. CS-UNR-101 (1999)
24. Daniel Lowd, P.D.: Naive bayes models for probability estimation. In: ICML '05 Proceedings of the 22nd International Conference on Machine Learning (2005)
25. Steve R.G.: Support vector machines for classification and regression (1998)
26. Popescu, P.S., Mihaescu, C.: Building an advanced dense classifier. In: Proceedings of the IISA2014 5-th International Conference on Information, Intelligence, Systems, and Applications, pp. 1–6. IEEE, Chania, Greece (2014)

Enhancing Machine Learning Models by Augmenting New Functionalities

P. S. Popescu⊙, M. C. Mihăescu, and M. L. Mocanu

Abstract The area of educational data mining is expansive, and machine learning models play a significant role in improving online educational environments. The e-Learning platforms previously used for distance learning and education tend to find their place in full-time education as an add-on for resources management and better interaction between users. This chapter tackles the problem of enhancing the machine learning models by offering methods for gathering more data that can be fitted in the actual datasets or building new datasets for specific tasks. The methods divide into three areas: messages, social media, and forum activities; each offers techniques for improving models by adding more data and more data and features. The experiments show that the gathered data can produce highly accurate models. The usefulness of such approaches is beyond improving the accuracy because these methods enable other functionalities like recommender systems that can be integrated (or enhanced if there are already) in the online educational environments.

Keywords Machine learning · Educational data mining · Decision trees

1 Introduction

Machine learning has become more and more common in many domains, and online education seems to need it and integrates it well. Using machine learning for education purposes helps teachers better understand the student, perform accurate forecasting or make accurate recommendations. These algorithms started to be used in online educational environments to lower the gap between these environments and full-time education, but once their utility was recognized, they tended to be used in more and more situations. The usage of e-Learning platforms is now extended in full-time education as it offers better resources management, improves the interaction between students, and integrates intelligent functionalities for knowledge tracking.

P. S. Popescu (✉) · M. C. Mihăescu · M. L. Mocanu
Department of Computer Science and Information Technology, University of Craiova, Str. A.I. Cuza, Nr. 13, Craiova, Romania
e-mail: stefan.popescu@edu.ucv.ro

© The Author(s), under exclusive license to Springer Nature Switzerland AG 2022
M. Mihaescu and M. Mihaescu (eds.), *Data Analytics in e-Learning: Approaches and Applications*, Intelligent Systems Reference Library 220,
https://doi.org/10.1007/978-3-030-96644-7_5

The methods presented in this chapter offer a solution for model improvement but from a different perspective which refers to the case of adding new functionalities or gathering different data from online educational environments to improve the machine learning models. One area referred to in this chapter is messaging activity which can provide more features for the dataset we use for model training. Gathering more data and features can increase the model accuracy and diversity, making it more generic and less prone to overfitting. Besides the benefits of the model enhancement, this data can be gathered from most of the e-Learning platform's databases, which is easy to use.

Another method presented in this chapter related to online educational environments refers to forum activity and user's interactions. Implementing such a forum helps students better interact with teachers and each other producing more data relevant to their activities. This extra data, which can be computed and gathered, can be used for specific models and to provide accurate recommendations.

All the functionalities presented in this chapter are essential and relevant for educational data mining as in the last years, there have been many efforts to develop better solutions for online education environments. This interdisciplinary domain is represented by research areas like information retrieval, visual data analytics, psych pedagogy, artificial intelligence, etc. Starting with user modelling research techniques in the early stages, the educational data mining research area included many contributions from artificial intelligence, intelligent tutoring systems, and technology-enhanced learning.

The research domain of Educational Data mining breakthrough in 2005 with a workshop referred to as 'Educational Data Mining' at AAAI'05-EDM conference in Pittsburg, USA [1]. After that, there were several related workshops, but the first annual dedicated international conference was three years later, in 2008 in Montreal [2].

In 2013 there were journals, conference papers, and the two books were published: the first is "Data mining in E-learning" [3] which is more oriented on online educational environments and has 17 chapters, and the next one is "Handbook of Educational Data Mining" [4] which is more comprehensible, have a total of 36 chapters and tackles different types of educational settings.

2   Related Work

The problem of predicting students' exam results was tackled in early 2003 Minaei-Bidgoli [5] presented a classification approach that modelled students to predict their exam grade based on features extracted from logged information from an online educational system. Later in 2010, Hamalainen and Vinni [6] wrote a paper called "Classifiers for Educational Data Mining", which was included in "Handbook of Educational Data Mining" in which they present a survey about the previous research on using data-driven classification for educational purposes, followed by a recall of the main principles affecting the model accuracy and finally giving several guidelines for accurate classification.

In 2012, a case study [7] described how to preprocess the data and extract the features for graduate students. They also presented results and discussions about applications of data mining techniques to graduate students' datasets.

A newer paper [8] approaches this problem of student modelling by using Dynamic Bayesian Networks. The authors also introduce a constrained optimization algorithm which can be used in order to learn the parameters of machine learning models and then they evaluate and clarify the accuracy of their method on five large-scale data sets extracted from different learning domains. As for validation, they use RMSE and AUC, and the results do not look bad, but it is not clear how they evaluate the ML models. Are they better than others? Even if it was published earlier, our approach was based on data gathered from users' messages. Other functionalities like computing predecessors or successors are based on the leaf in which an instance is located. In this way, the data analyst can navigate through a very dense tree.

2.1 Related Work in Student Modelling Based on Text Analysis

Text analysis, and more precisely, messages analysis, is an essential factor for Educational Data Mining offering important information and helping instructors better understand the learning behaviour. This type of analysis performed on e-Learning systems started a long time ago and is actual even these days. In 2003 Mochizuki et al. [9] published a paper examining a method for visualizing discussions by focusing on the relationship between topic keywords and each student using a correspondence analysis as described in [10]. Even if it was very early for this domain, their proposed method was presenting a map between topic parts from forums and students' commitment in the learning environment, according to interviews in which students took part. The map aimed to help students engage in active discussions by providing a reflection upon their interests and enhancing collaborative learning [11].

Another newer paper that tackled messages analysis and text mining is [12], which had the purpose of doing a threefold analysis of the data related to students' participation in the online forums offered by their faculties. The first analysis focuses on how text analysis techniques can efficiently analyze the content of the messages posted in these forums. The second analysis focuses on the fact that the network of students interacting using a forum can be adequately processed based on social network analysis techniques. Finally, the combined knowledge from the previously presented techniques can provide teachers with practical and valuable information for understanding and evaluating the learning process, especially in an online learning environment. Their study was conducted by using data gathered from the online forums of the Hellenic Open University (HOU).

Later in this area, many papers like [13, 14] used features extracted from messages to enhance or build machine learning models used in online educational environments. Nevertheless, before these two papers, one published by Baker et al. [15] described a method for faster and more precise labelling of student behaviour. Their method involved hierarchical classifiers such as Latent Response Models (LRMs) [16] and non-hierarchical classifiers such as Decision Trees and generated labels that can further train classifiers based on students' behaviour. They used this method to label data within 20 intelligent tutor units on Algebra. They also found that text replays are even six times faster for each label than previous methods for automatic generating labels. However, generating classifiers on retrospective data at the coder's convenience makes this technique 40 times faster than other quantitative observation methods.

A newer paper published in 2019 [17] focuses on answering three critical questions: Which text mining techniques are *primarily used in educational environments? Which are the most used educational resources? Furthermore, which are the main applications or educational goals?* The authors found an explosive growth of distance learning environments which generates a big volume of data, especially in .txt format from several sources like chats, forums, social networks, assessments, essays, etc. This increase generates exciting challenges on how to mine the data to provide helpful knowledge for scholastic participants. Despite the growing number of educational data mining applications of text mining published recently, they have not found any survey that tackles this direction, and the paper presents a systematic overview of the current status of the Educational Text Mining research area, presenting even very new papers like [18–20] which are very new and validate the actual status of this research area.

Another new and more technical paper is [21], which focuses on analyzing social interactions within online discussion forums, and the proposed system has two significant steps: 1 which is for Learners' vector building and 2 which is for SVM-based classifier. Their experiments present the efficacy of the proposed system achieving a great accuracy of 0.9 and a Cohen's $K = 0.89$. Their research aims to ensure good scaffolding by offering tutors the chance to observe learners' cognitive activities, especially their cognitive engagement. Based on the gathered data, they present a system for classifying students according to their stages of cognitive engagement.

3 Design of Improved User Modelling Based on Messages from E-Learning Platforms

Learner's profiling is, in many cases, a complex task and modelling a user based on its messaging activity with the other users can be achieved in several ways one being by using a graph data structure. For our method, the nodes represents students or professors, and the vertices suggest the interactions created between two entities along with their strengths.

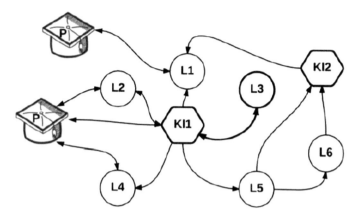

Fig. 1 Graph example

In Fig. 1, we present a part of a graph that can be extracted from an online educational environment. There are three entities defined here: L, a regular student (learner), KL, a Key student (Learner), and P, which is a professor. The relation between them may belong to one of the following types: unidirectional, bidirectional and robust bi-directional. Unidirectional relations between two actors correspond to message that were sent but did not receive a response. This might indicate that the initial message may have been an statement that did not ask for a reply or a regular message that did not respond for any reasons. The bidirectional relations refer to users with a low interaction (i.e., they exchange a few messages). The last type of relation is the solid bidirectional one. This is relevant because it marks some intensive interaction and changing messages like chatting. This strong relation can be beneficial in achieving essential students. In many cases, if we want to search entities that act as outliers in dataset, the "Interquartile Range" filtering algorithm can be useful and a good choice.

The first challenge for this method is to build a list of attributes extracted from an online learning environment (e-Learning platform) related to the learner's messaging activity. Obtained attributes are divided into two categories: student-oriented and message-oriented.

The student-oriented list of features

General features regarding a student can be gathered from every message. From this category, we can compute the following attributes:

- *MSG_TOTAL_REC_COL* → the total number of messages received on a conversation initiated by them from colleagues
- *MSG_TOTAL_SENT_COL* → the total number of messages sent on a conversation initiated by them to colleagues
- *MSG_TOTAL_REC_PROF* → the total number of messages received on a conversation initiated by them from academic staff

- *MSG_TOTAL_SENT_PROF* → the total number of messages sent on a conversation initiated by them to academic staff
- *MSG_TOTAL_ANS_REC* → the total number of messages a got by a user from responders
- *MSG_TOTAL_ANS_SEND* → the total number of send messages as a response to other messages
- *MSG_MEAN_RECV* → average delay between received messages in minutes
- *MSG_MEAN_SEND* → average delay between sending messages in minutes
- *MSG_MEAN_ANSWER* → average delay when answering a message
- *MSG_NOT_ANS* → messages were sent, but no reply was given. This feature is essential because some users may try to find answers to several problems without solving them. These kinds of users may be treated as outliers in data.

List of features oriented on messages

- *MSG_AUDIENCE* → the number of students that receives a message
- *MSG_TYPE* → represents the type of message (public or private)
- *MSG_LENGHT* → length of the message in number of characters
- *MSG_COURSE_RELEVANT* → indicates the message if it is related to a course subject
- *MSG_SOCIAL* → indicates if a message is social or focused on a subject
- *MSG_LINK* → marks if the message contains an external or internal link.
- *MSG_TOTALRECV_WDS* → the total counted words from the received messages
- *MSG_TOTALSEND_WDS* → average delay between received messages
- *MSG_TOTALRECV_KWS* → the total counted keywords from received messages
- *MSG_TOTALSEND_KWS* → the total counted keywords (concepts) from sent messages
- *MSG_RECVPERC_KWS* → the percentage of keywords from counted words from received messages
- *MSG_SENDPERC_KWS* → the percentage of keywords from counted words from sent messages.

3.1 Algorithm Selection for Data Analysis

Several classification algorithms are used in this educational research area. For our method, we will use some of the most relevant and useful ones to validate our findings. We chose these algorithms because they can be easily of trained and tested on our data. The logging system is designed to run for at least a year and, starting from that time, to gather data that includes the students' scores. This logged data is converted in a friendly format and used for algorithm training. Current dataset is collected at several predefined checkpoints, and on each checkpoint, the algorithm will be trained and validated on the dataset, afterwards, predictions will be made. As soon as the algorithm predicts a student's academic failure, the system will alert and send him a warning message.

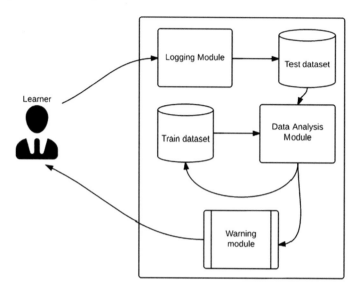

Fig. 2 Data pipeline overview

In Fig. 2, we have the whole system's overview. First, we need to mention that the training dataset was constructed from a couple of years, and the presented modules are employed only during specific checkpoints. The logging module collects data from the users and completes the test dataset. Based on the training dataset, the "Data Analysis Module" may forecast any failure triggering alarm for any instance in the test dataset. The failure (warning module) module will decide how the messages will be sent, acting as a task scheduler, and the student will be aware of his status.

The chosen machine learning algorithms are from each main category: Decision Trees, Functions, lazy, Naive Bayes, and rule based classifier. We will evaluate the results on the proposed data set will reveal which outperforms others and which performs best for this specific situation.

This contribution is presented in detail in the paper "Messaging activity impact on student's profiling" [22], and for proof purposes, we have chosen a dataset including only 98 instances collected from the Tesys e-Learning platform. These instances are relevant to the last academic year and represent the training data used as input for the algorithms. One significant feature that needs to be mentioned is that in our learning programs, the number of students will not exceed 120, and it is not relevant to create a dataset from two different learning programs because they will not have enough similar courses and their messaging topics may not coincide.

Figure 3 is presents a sample from the experimental dataset used for training. In the figure, we can first observe the features that describe an instance. Not all of the features listed in this paper were used due to the limitations of the e-learning platform. Their meaning is presented in the subsection above.

```
@RELATION messages

@ATTRIBUTE MSG_TOTAL_SENT_COL numeric
@ATTRIBUTE MSG_TOTAL_REC_COL numeric
@ATTRIBUTE MSG_TOTAL_SENT_PROF numeric
@ATTRIBUTE MSG_TOTAL_REC_PROF numeric
@ATTRIBUTE MSG_TOTAL_ANS_SEND numeric
@ATTRIBUTE MSG_TOTAL_ANS_REC numeric
@ATTRIBUTE MSG_MEAN_SEND numeric
@ATTRIBUTE MSG_MEAN_RECV numeric
@ATTRIBUTE MSG_MEAN_ANSWER numeric
@ATTRIBUTE MSG_NOT_ANS numeric
@ATTRIBUTE class {fail,succeed}

@DATA
138,90,1,1,78,127,10.00,10.00,251.67,30,succeed
111,73,0,0,93,140,11.00,10.00,247.98,21,fail
92,55,1,1,61,104,15.00,14.00,269.96,25,fail
122,78,0,0,90,149,10.00,10.00,254.98,26,fail
119,67,4,2,76,128,11.00,11.00,241.28,33,succeed
99,58,6,3,81,131,12.00,11.00,263.60,26,succeed
114,61,2,1,59,98,13.00,14.00,262.59,35,succeed
103,64,1,1,113,167,10.00,9.00,264.45,30,fail
```

Fig. 3 Sample dataset

For validation purposes, we performed some experiments for the main categories of machine learning algorithms. In the tables presented below, we present the algorithm's accuracy for each algorithm.

Table 1 are presents the experiments performed on decision Trees. As we can see here, we have seven decision tree algorithms, each of them with the obtained accuracy. The best performing algorithm from the decision trees category is RandomForest (85.7143), and the algorithm with the lowest accuracy is HoeffdingTree (73.4694). J48, one of the most common decision trees, scores average with an accuracy equal to 82.6531.

Bayes Classification algorithms are a set of simple probabilistic algorithms based on applying Bayes' theorem. In Table 2, we present the main Bayes algorithms available in Weka and their accuracy based on the dataset. For this dataset, only BayesNet offers an average accuracy, the others just slightly better than HoeffdingTree.

Table 1 Set of decision trees algorithms used for experiments and results

Algorithm	Accuracy
J48	82.6531
RandomForest	85.7143
LMT	80.6122
DecisionStump	82.6531
HoeffdingTree	73.4694

Table 2 Set of Bayes classifiers used

Algorithm	Accuracy
BayesNet	82.6531
NaiveBayes	74.4898
NaiveBayesMultinomial	74.4898
NaiveBayesUpdateable	74.4898

Table 3 presents the most critical "Function" algorithms. These algorithms' performance is also average for this dataset, and some of them are even lower than the lowest decision tree.

Rule-Based classifiers are always compared with decision trees because these algorithms are based on rules generated for every instance. Table 4 are presented the most critical Rule-Based algorithms from Weka. As we can see, their accuracy is above average if we exclude ZeroR.

In Table 5, the last one, we have a comparison between lazy classifiers. LWL performs above average, but the other two have the lowest values.

Table 3 Set of function classifiers used

Algorithms	Accuracy
Logistic	80.6122
MultilayerPerceptron	71.4286
SimpleLogistic	80.6122
VotedPerceptron	76.5306
SGD	80.6122

Table 4 Set of rule-based classifiers

Algorithms	Accuracy
ZeroR	76.5306
OneR	83.6735
DecisionTable	81.6327
Jrip	81.6327

Table 5 Set of Lazy classifiers	Algorithms	Accuracy
	KStar	64.2857
	IBk	67.3469
	LWL	82.6531

Based on the tables presented above, we can say that if we want to perform data analysis on a dataset that has a format similar to the one presented in this paper, we should use Decision Trees for sure.

3.2 Design of New Functionalities to Improve Student Modelling Based on Forum Activity

Building and integrating a *SmartForum* in e-Learning platform can help users achieve several goals like offering useful functionalities for enhancing the students' performance; create a strong environment for improving the users' interaction and generate high-quality logged data that can be analyzed and used for training algorithms. There is a previous publication [23] which further explain this contribution and can be consulted for more complete information regarding the design procedure along with detailed examples.

Improving the users' collaboration and interaction can be achieved by expanding the method of interaction by number and also quality. Forums are virtual environments that make available message exchange and conversation. A particularity of forums is that the messages are primarily public or accessible at least for registered users; another particularity is that the messages are classified in topics of conversation, so the data is well structured. In case of this system, messages written on *SmartForum* are available for the actors from a specific course which in our case are the students and professors. The benefit of implement an extra mean of interaction which has specific particularities can increase students' engagement and its impact can be measured using specific HCI metrics. In order to validate the system, we need to have it fully integrated and used for at least an academic semester, so we can estimate the level of interaction. Based on the evaluation we can set several parameters like several notifications frequency or trigger messages of interest which can have a specific rate.

The educational data mining research area is vast, and it encapsulates plenty of data sources but the data which is already logged in online educational environments refer users and the learning resources uploaded in the platforms. In our approach we consider Smart Forum as a module for an e-Learning platform that can produce an additional amount of data which involves other learning resources and increases the user's interaction. Based on the newly produced data we can analyze what sections are more debated which can give us an insight regarding the student's engagement and what concepts from the courses are referred mostly in the forum. We can also improve the learner's profiling by adding more data based on the activities that can be performed on the forum and later gathered from it. If we manage to log more data, we can reveal new patterns and also if we manage to add more attributes, we can improve the actual machine learning models.

Another relevant topic of interest in the educational data mining research area is bringing new functionalities to the existing system and also improving the classification models. At this moment, there are several existing solutions, but there is also plenty of space for improvement and progress. Another benefit is that if we manage to collect more data and use it to improve the machine learning model can bring more accurate results. There are also associations which can be mined between forum activity and activity through the e-Learning platform like using the learning resources or taking tests. Other correlations can be explored between the forum and the grades obtained at tests or final exams. One matter that may influence the student's results is their interaction and what they can be taught. Using our described e-Learning platform, a study can discover if students who interact more with colleagues through virtual environments and learn more from other students will achieve better results or at least have an ascending learning curve.

The *SmartForum* creates an online environment for interaction between all people, not only between students, who perform online educational environments like professors, students, and administrative staff. Every person who uses this forum has a set of specific tasks associated with their topics of interest: students can discuss the subjects related to the courses and ask questions, professors may respond to questions and may launch challenges regarding specific subjects within a course, and finally the administrative staff may respond to questions of common interest or make announces which can be interesting for most of the users.

Figure 4 presents how the courses and modules (or sections) are distributed over the Tesys e-Learning platform's infrastructures. The setup of the e-Learning platform preset here has 3 (but it can be more, depending on the faculty setup) years of study; each year of study have several sections (study programs) or modules. These sections which in our platform are called modules have the letter "M" in their node, and every study program has a set of courses attached to it represented by the letter "D". We

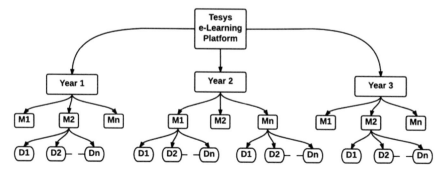

Fig. 4 Tesys components distribution

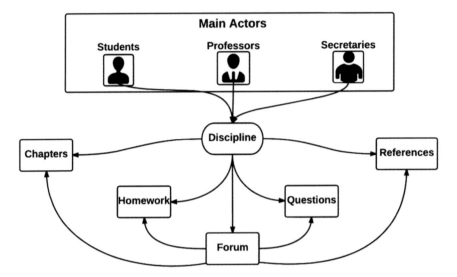

Fig. 5 Forum integration

added in the figure courses only on some modules Because of the space limitations, but there are usually a set of around six disciplines on each module. This figure is relevant because in shows how the smart forum is connected to every course being well distributed across modules in the platform. The forum distribution over the courses ensures that the topics discussed in the forum are strongly connected to the course chapters used to set the discipline and ensures that the information from the forum is relevant to the courses.

Figure 5 presents the components that are encapsulated in a discipline. Starting from the upper part of the picture, the main actors perform their activities in one discipline: students, secretaries and professors. In this case we exclude the administrative and setup role because this one is not used in day-by-day activities. Every actor has his interface and access to specific features and modules of the e-Learning platform depending on their level of access. Homework, questions, chapters, and external references are placed at the same level as the forum and are related to the discipline covering several learning areas. We also define chapters which are represented by the documents uploaded for each course and homework is used to assign students tasks that need more time and involve reading other materials while questions are used for exams and tests and usually in any of the exams or tests setups the answer needs to be provided fast. The forum is used to discuss subjects related to a specific discipline; this makes the data stored in it more focused on several subjects. For example, assume a data analyst wants to analyze the subjects related to course "x", he wants to find the specific topics related to the course "x" by using some analysis techniques which will most likely not provide 100% accuracy, or the subjects will have tags that were assigned by the subject initiator which will also not provide 100% accuracy all the time. Distributing the forum over the course offers one more topic

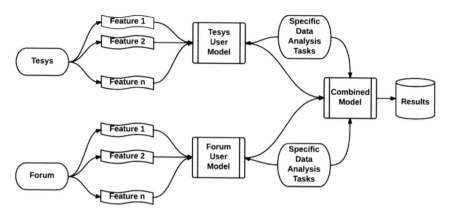

Fig. 6 Data analysis pipeline

focus related feature over the existing ones. There will always be online moderators who are represented by professors or/and top students; these moderators will make sure that there are no topics out of the course area on a specific forum.

Figure 6 presents the proposed data analysis pipeline. There are two main components that contribute to data logging: Smart Forum and Tesys which are presented in Fig. 6. Although the *SmartForum* is integrated as a component into the e-learning platform, it provides a different amount of data that can be used as features and can contribute to the existing data we collect. This new data can be combined or used as stand-alone with the features extracted from Tesys to build better models or to improve the existing ones. There are specific data analysis tasks that will be performed only on the data extracted from Tesys or from the Smart Forum, and there will be several specific tasks that will be performed only on the joint model.

Validation of the system in terms of data analysis can be performed by comparing the results obtained using the joint models with the models obtained using only the features from Tesys and this way we can evaluate the impact of the additional features added.

Forum functionalities

In this forum, there are two sorts of actions available: the regular ones which are common for any classical forum and smart ones that include the usage of several data mining and machine learning algorithms.

There are several attributes that may be extracted using Smart Forum as it is designed now. These attributes enable better user modelling based on the forum activity analysis, but this list of features is not exhaustive and can be further improved.

- *NO_POSTS*—the whole count of posts created in the forum
- *NO_TOPICS*—the whole count of topics started in the forum
- *NO_QUESTIONS*—the whole count of questions asked in the forum
- *NO_USEFUL_ANSWER*—the whole count of answers contributed on the forum marked as helpful by users who read the message

- *NO_ACTIVE_DAYS _FORUM*—the whole count of days when the user read or posted in the forum
- *AVG_TIME_READ*—the mean time that passed between a new post was made by a user and the moment when the user read that post. This can be also an indicative of user's engagement
- *NO_CONCEPTS_COURSE*—the whole number of concepts from a course that a student in his posts referred to. This attribute will give us a point of view of how focused on the course is a student when he posts
- *FRQ_POSTS*—the frequency of posting new messages which is computed in as the number of posts divided by the number of days active in the forum
- *FRQ_TOPICS*—the frequency of creating new topics which is computed in as the number of posts divided by the number of days active in the forum.

We can model the users and their activity based on these attributes, but only after performing experiments can we say which attributes are relevant and useful for model improvements or even creating a specific model. Attribute selection is a generally complex challenge because we need to choose the features that are strongly related to the learner's activity but also, we need to choose the ones that offer better results for the chosen algorithms.

Main Forum Functionalities

In this feature, we include the possibility to add, delete or edit a message for a specific amount of time. It is also essential to save the original message in the database for a period so that the moderators can solve any complaints.

User profiles information section and provide the number of messages posted and topics started on the forum for the current discipline and even the whole number of posts made on the forum. The profiles also refer to the whole number of answered questions and the number of answers which were marked as correct and useful by other users. Still, concerning the user profile information, we may have several questions from the forum that were answered by the user and was marked as a helpful answer by other users, and this can be used to increase the reputation of that user. This capability provides us with a vision of how trustworthy the student is when he answers questions from the forum.

Smart Functionalities

The smart functionalities make the distinction between a standard discussion forum which can be easily added in an e-Learning platform and our proposed *SmartForum*. These functionalities are supported by some data mining and machine learning algorithms and are based on the data which can be collected from the forum. Gathering several features which can be used for training the machine learning algorithms from the proposed *SmartForum* creates the possibility of implementing the functionalities that make the forum get the "*smart*" naming and characteristics. Below we describe the main functionalities that can be developed along with a short description.

Regarding the educational area, the *SmartForum* can offer several subjects of interest for students and increase their knowledge. This functionality will be accomplished by analyzing the learner's answered questions and text analysis over their posts. If we can match between the concepts that can be gathered from the wrong answered questions and the concepts gathered from some specific posts or topics from the smart forum, we can recommend those posts to the learners. Going further for this feature, we can also see what concepts are considered attractive by the student, and we can recommend several topics that were not read by him.

This capability can also be accomplished by matching concepts gathered from the forum and the concepts gathered from the topics that were read or answered by the student, and the correctly answered questions.

There are two ways to gain the subjects that may be relevant for a student: offering the ones that are referred a lot or the ones that reveal lower results. In order to get the subjects that are the most talked about, we need to perform text analysis and classification (i.e., concept extraction) on both the messages posted on the forum and on questions answered.

Figure 7 presents an overview of the proposed system. First, we have to use clustering algorithms (i.e., SKM clustering algorithm) to get the learning addressed mainly by the students and the area from the forum that is most addressed. Then, the concepts can be gathered using stemming algorithms from both collections. Having a more significant percentage of match concepts means a better matching, and then we can recommend them to be read.

In order to improve student's academic performances, we need to obtain the learning resources (questions and Homework) on which the learner got poorer results, then use stemming from getting the familiar concepts from them. Based on posts analysis, we can obtain the most related messages and offer them to students for reading.

Computing trends for learners will also be a smart feature available in Smart Forum. This feature is referred to by generating charts that will present how the number of posts, number of answered questions that were marked (id available), number of questions raised, or number of topics read fluctuate over a while. This can

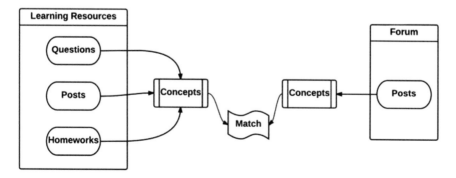

Fig. 7 Overview of the designed mechanism

be useful for measuring and predicting the student's engagement over the platform and, more specifically, over the smart forum.

Figure 8 presents a trend example computed for a student over 12 weeks which, in our case, represents the entire semester. On the OX axis, we have marked the weeks starting from 1 and going to a maximum of 12, and on the OY axis we plot the student's amount of activity. At a more profound analysis, we can observe that the student starts from an activity value of 4.5 and don't have much progress until he reaches week six, then there is slight activity decrease which affects the next weeks, and then the student obtains a rising trend until he reaches the last week of the whole semester. On the chart, there is also the green line with no dots on it which represents the trend that can be computed based on the student's activity. As we can observe in Fig. 8 in this case, the green line indicates a rising trend based on the student's activity. This proposed activity trend can be approximated using several attributes which can be gathered from the Smart Forum and the accuracy of the result is dependent on both the amount of data we can collect and also on how relevant this data is for computing the student's activity.

Going further with the analysis, Fig. 9 presents the same activity from Fig. 8 but discretized for three periods; on the OX axis we define the weeks which are from 1 to 12 like in the previous figure, and on OY on each graph, we have the activity level. Similar to Fig. 8, the grey line with dots represents the level of student's activity, the green line represents the detected ascending trends, the steady trend is represented by the blue line, and the red line represents a downward trend computed for the student proposed for analysis.

In Fig. 9 there are two subfigures and for (a), the first phase of the activity is represented by the first three weeks of the semester and have an ascending trend.

Fig. 8 Trend example

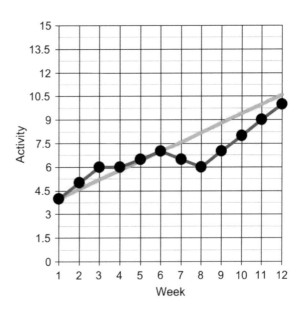

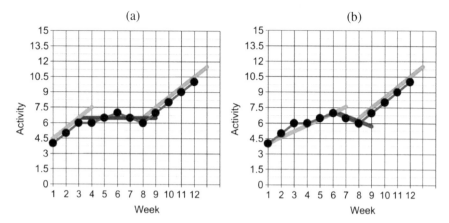

Fig. 9 The trend discretized for three periods

The second phase starts at week four and goes to week 9 as a steady trend as can be observed in the figure. At week 9 there is another trend switch which goes ascending until the semester is over.

Based on the same chart, there is another approach for computing trends as presented in (b) where we assess that the first six weeks represent a rising trend, followed by a falling trend which starts at week 6 and goes to week 8, and then we go back to a trend which keeps rising until the end of the semester. This approach may be better and simpler to estimate but it is hard to say which one fits better for that situation.

Estimating trends is an important task as it can reveal the student's evolution but also needs a fine tuning setting the best period for computing a trend. Finding the best dimension for selecting a trend window is crucial if we can to trigger alerts if a student enters a risky situation. Based on the comparison of the two subsections from Fig. 8 we can see that having the same trend we can have two different situations which can imply different actions because if we consider the situation from (a) there is no need for an alarm comparing to the situation from (b) when we have a short descending trend. Based on this situation is also important to detect the trends early in order to be able to take fast decisions.

Generating alerts during the whole semester regarding the student's failure is another feature of the forum that needs to be considered. First, several tests need to be performed to see if the student's activity on the forum will forecast its success/failure or even his final exam grade.

4 Conclusions

Machine learning models can be optimized in various ways, and most of them are already addressed in the existing literature. Model optimization was well addressed in the specific literature but primarily focused on dataset preparation, parameters tuning, and algorithm selection. These approaches can improve the result/score, but there is no significant change in most cases, and there is also a chance to create overfitting when applying them.

The changes proposed in this chapter focus mainly on implementing functionalities that gather and/or produce more data to create models that focus better on students' actual needs. Using messages to extract extra information can improve the dataset and the model's accuracy, as stated in the experiments. The dataset extracted can be used stand alone and integrated into another dataset along with other features that offer great flexibility.

Implementing a smart forum as an addition to a classical e-Learning platform leads to possibilities for enhancing the datasets and models. As presented above, there are plenty of forecasting possibilities and trends which can be computed. One thing that needs to be mentioned is that these methods are pretty generic and can be implemented in almost any e-Learning platform.

As future work, we plan to investigate other modalities to enhance machine learning models and proceed further in exploring the possibilities for deep learning models as they tend to be more and more popular. As the current context of working and learning from home tends to be more and more used, functionalities like automatic student detection or automatic cheat on tests detection can also improve the systems and give educators a better insight into the student's performance.

References

1. Beck, J.: Proceedings of AAAI2005 Workshop on Educational Data Mining (2005)
2. Baker, R.B.: Educational data mining 2008. In: 1st International Conference on Educational Data Mining, Proceedings. Montreal, Quebec, Canada (2008)
3. Romero, C., Ventura, S.: Data Mining in E-Learning. WIT Press (2006)
4. Ventura, C.R.: Handbook of Educational Data Mining. CRC Press (2010)
5. Minaei-Bidgoli, B., Kashy, D.: Predicting student performance: an application of data mining methods with an educational Web-based system. Front. Educ. (2003)
6. Vinni, W.H.: Classifiers for educational data mining. In: Romero, S.V.C. (ed.) Handbook of Educational Data Mining (2010)
7. AbuTair, M.M., Alaa, M.-H.: Mining educational data to improve students' performance: a case study. Int. J. Inform. Commun. Technol. Res. (2012)
8. Kaeser, T.K.: Dynamic bayesian networks for student modeling. IEEE Trans. Learn. Technol. (2017)
9. Mochizuki, T., Fujitani, S., Isshiki, Y., Yamauchi, Y., Kato, H.: Assessment of collaborative learning for students: making the state of discussion visible for their reflection by text mining of electronic forums. In: E-Learn: World Conference on E-Learning in Corporate, Government, Healthcare, and Higher Education, pp. 285–288. Association for the Advancement of Computing in Education (AACE) (2003)

10. Fujitani, S., Mochizuki, T., Kato, H., Isshiki, Y., Yamauchi, Y.: Development of collaborative learning assessment tool with multivariate analysis applied to electronic discussion forums. In: E-Learn: world Conference on E-Learning in Corporate, Government, Healthcare, and Higher Education, pp. 200–203. Association for the Advancement of Computing in Education (AACE) (2003)
11. Ueno, M.: Data mining and text mining technologies for collaborative learning in an ILMS "Ssamurai". In: IEEE International Conference on Advanced Learning Technologies, 2004. Proceedings, pp. 1052–1053. IEEE (2004)
12. Lotsari, E., Verykios, V.S., Panagiotakopoulos, C., Kalles, D.: A learning analytics methodology for student profiling. In: Hellenic Conference on Artificial Intelligence, pp. 300–312. Springer, Cham (2014)
13. Romero, C., Romero, J.R., Luna, J. M., Ventura, S.: Mining rare association rules from e-learning data. In: Educational Data Mining (2010)
14. Romero, C., Ventura, S., Vasilyeva, E., Pechenizkiy, M.: Class association rules mining from students' test data. In: Educational Data Mining (2010)
15. Baker, R., de Carvalho, A. Labeling student behavior faster and more precisely with text replays. In: Educational Data Mining (2008)
16. Khajah, M., Wing, R., Lindsey, R., Mozer, M.: Integrating latent-factor and knowledge-tracing models to predict individual differences in learning. In: Educational Data Mining 2014 (2008)
17. Ferreira-Mello, R., André, M., Pinheiro, A., Costa, E., Romero, C.: Text mining in education. Wiley Interdiscip. Rev.: Data Min. Knowl. Discov. **9**(6), e1332 (2019)
18. Yildiz, T.K., Atagün, E., Bayiroğlu, H., Timuçin, T., Gündüz, H.: Text mining based decision making process in Kickstarter Platform. In: The International Conference on Artificial Intelligence and Applied Mathematics in Engineering, pp. 344–349. Springer, Cham (2020)
19. Kansal, A.K., Gautam, J., Chintalapudi, N., Jain, S., Battineni, G.: Google trend analysis and paradigm shift of online education platforms during the COVID-19 pandemic. Infect. Dis. Rep. **13**(2), 418–428 (2021)
20. Pans Sancho, M.À., Madera Gil, J., González Moreno, L.M., Pellicer Chenoll, M.T.: Physical activity and exercise: text mining analysis. Int. J. Environ. Res. Public Health **18**(18), 9642 (2021)
21. Hayati, H., Idrissi, M.K., Bennani, S.: Automatic classification for cognitive engagement in online discussion forums: text mining and machine learning approach. In: International Conference on Artificial Intelligence in Education, pp. 114–118. Springer, Cham (2020)
22. Popescu, P.Ş., Mocanu, M., Burdescu, D.D., Mihăescu, M.C.: Messaging activity impact on learner's profiling. In: 2015 6th International Conference on Information, Intelligence, Systems and Applications (IISA), pp. 1–6. IEEE (2015)
23. Popescu, P.Ş., Mocanu, M., Ionaşcu, C., Mihăescu, M.C.: Design of an advanced smart forum for Tesys e-learning platform. In: IFIP International Conference on Artificial Intelligence Applications and Innovations, pp. 305–316. Springer, Cham (2016)

Increasing Engagement in e-Learning Systems

M. L. Mocanu⊙, P. S. Popescu, C. M. Ionaşcu, and M. C. Mihăescu

Abstract Recent research has shown the importance of social activities in e-learning processes and revealed their effect on academic achievement. Even within innovative e-learning models, the social dimensions of engagement have a long time been neglected. This chapter aims to analyze the activities performed by students in different social applications (blog, tweet, and Wiki) and understand their possible relation with their engagement and academic results. Experimental results show a correlation between social activities, indicating engagement, and academic performance. Furthermore, the lack of participation during the interactive activities may indicate insufficient preparation in performing the learning tasks. Still, we provided a data analysis pipeline that finds students with significant engagement but with low academic performance and students with low engagement but with high academic performance. Proper analysis of various categories of students opens the way towards an accurate classification of students in terms of their engagement and academic proficiency.

Keywords Machine learning · Educational data mining · User modelling

1 Introduction

In the last years, many efforts have been made towards developing better solutions for online education environments. The e-Learning systems research area began to include many contributions from various domains, such as intelligent tutoring systems (ITS), human–computer interfaces (HCI), social networks and social media, and technology-enhanced learning. This chapter details the relationship between social dimensions of engagement, participation, and academic performance.

M. L. Mocanu (✉) · P. S. Popescu · M. C. Mihăescu
Department of Computer Science and Information Technology, University of Craiova, Str. A.I. Cuza, Nr. 13, Craiova, Romania
e-mail: mihai.mocanu@edu.ucv.ro

C. M. Ionaşcu
Department of Statistics and Economic Informatics, University of Craiova, Str. A.I. Cuza, Nr. 13, Craiova, Romania

© The Author(s), under exclusive license to Springer Nature Switzerland AG 2022
M. Mihaescu and M. Mihaescu (eds.), *Data Analytics in e-Learning: Approaches and Applications*, Intelligent Systems Reference Library 220,
https://doi.org/10.1007/978-3-030-96644-7_6

Analyzing social interactions in an e-Learning context can be challenging as a marker of engagement. Without direct interactions of instructors and learners, e-Learning systems lack the human learning social dimension, often limiting learners' engagement. The HCI may also contain critical functionalities that need to be fully understood by the users, like testing procedures or taking exams. Another aspect that is not discussed in this chapter is that a low-quality interface may consume a significant amount of time and reduce engagement from the learners when taking tests or an exam which may lead to low results.

When addressing HCI as a research area, we also need to consider *the goals* we want to achieve (in this case, they are related to *increasing learners' engagement*) and *the context* of the study. The goals are highly dependent on the context of the study as not any goal can be achieved in any context. Mainly when we define HCI related goals, we refer to the user's possibility to anticipate the actions they must perform to produce an effective, efficient, and safe interaction and use the best practices to build a system that is enough for users. The context of the study is more complex as it is influenced by many factors: organizational, environmental, comfort, or task factor. The user, constraints, system functionality, productivity, and system interface, are other factors that influence the actions that need to be performed to evaluate and optimize an application. In order to produce an efficient evaluation of the target system and provide valuable feedback, we need to identify the context and the factors that influence the design.

The popularity and impact of social media activity have been continuously expanding worldwide [1], and a similar trend is visible in education [2, 3]. A large body of research reported using blogs, wikis, microblogging tools, social bookmarking tools or social networking services for various educational purposes, with different instructional scenarios and disciplines of study [4]. Therefore, it is essential to evaluate the use of social media for e-Learning and to study the learning processes in social media.

Research reports indicate that the usage of social media tools generally increases participation and engagement. Learning performance indicators are an integral part of the educational process and help teachers provide appropriate feedback and create proper modes of evaluation or intervention [5]. The effect of active participation on student performance has been widely investigated in online learning settings already from some time ago [6–8]. The results of performed studies are sometimes contradictory [9] regarding how online engagement explains academic performance. Several studies found that online participation is a reliable indicator of student performance [10] and improves learning effectiveness [11]. Still, other studies concluded that students learned equally well regardless of their level of online participation [12].

The classical mixed-effects analysis of variance (ANOVA) models has been used by [13] to evaluate the impact on learning outcomes induced by using Twitter. The National Survey of Student Engagement instrument has been used to measure engagement. The results showed a significant increase in grades and engagement for students who used Twitter for various academic discussions.

A collaborative wiki tool has been used by [14] to investigate students' engagement and predict academic performance. There were used Wiki activity indicators such as the number of page edits, the number of different articles edited, the number of days on which the student-edited the Wiki. There was a significant correlation (but without proving causation) between wiki activity indicators and final grade, irrespective of whether the grades were low or high. Still, students that used Wiki obtained an increase of 5% points in grade.

The simultaneous usage of several social media tools and evaluating their impact on learning performance has been investigated by [9]. Using Principal Component Analysis (PCA), a subset of student activities relevant to building up an orthogonal representation space may be identified. The identified PCs may represent the most appropriate activities to predict the final grade. Results showed that the number of blog posts, wiki page revisions, and shared bookmarks are reliable predictors of student performance.

The proposed approach described in this chapter uses a multiple linear regression approach to investigate the relationship between students' activity on social media tools and their final grades. The motivation for this approach is sustained by the data types of independent and dependent variables. The independent variables are numeric (i.e., counts and frequencies of social interactions), and the dependent variable is also numeric as it is the final grade.

2 Proposed Approaches for Increasing Engagement

2.1 Modelling Students Based on Their Activity on Social Media Platforms

The first step in student modelling uses the data gathered from BlogSpot, Twitter, and MediaWiki, and builds user-generated student models [15]. The context of the study is represented by the undergraduate Web Application Development (WAD) course at the Computer Science and Information Technology department at the University of Craiova. The practical activity at WAD consists of a relatively complex project which the students develop during the semester.

Students are paired in teams of 3–4, and their task consists of implementing the application while using the social media tools for collaboration. The tasks, as well as organizing information and resources, is performed by using Media Wiki. Another tool that has been intensively used is Blogger. Each team uses this tool for monitoring the progress of the project, the key ideas and resources and for feedback and answers to the problems raised by the peers. Finally, Twitter has been used for connecting peers by posing short questions, news or announcements, and possibly status updates regarding the project.

eMuse platform [16] has been used for several years to monitor students' activity on BlogSpot, Twitter, and MediaWiki, although minor improvements have been implemented according to students' and professors' feedback. 66 and 53 students participated in the study in the first two years, respectively.

The first tool that has been developed is a custom *.arff* builder, which takes the raw data produced by the social media tool and creates a structured data file that is ready for analysis. Once the data is prepared for analysis, we use our ranker [17] and linear regression from Weka to obtain results. The structured data file consists of all computed features for students who performed activities using the social media tools. The *.arff* file is ready for processing by Weka algorithms since it follows the specifications for data representation. The ranker builds the training dataset in a semi-automatic way. It orders the classes of students in terms of their performed activities. It also allows the analysis to be completed in a highly interactive and visual manner while presenting the instances within the data model. The ranker correctly visualizes the rules corresponding to all leaves (i.e., classes). Finally, the multiple linear regression from Weka is integrated with the *.arff* builder tool and ranker in a pipeline that processes the dataset and builds the data model. Setting up the data analysis pipeline is configurable, and model rebuilding is semi-automatic. Quality metrics are available immediately such that interpretation of results may be accomplished quickly. Figure 1 presents the structure of the data analysis pipeline with principal components and their interactions and results.

The first step is represented by defining the features that characterize a student. The custom *.arff* builder takes the raw learner traces as input from the social media tools and builds the dataset. The eMUSE database actions are processed so that each student gets its feature values computed. The following attributes are computed:

id_action—the unique identifier of the performed action
id_student—the identifier of the student who performed the action
id_tool—the identifier of the used social media tool
action_name—the specific type of action that has been performed
action_description—details about the action
date—the timestamp of the action.

Additionally, the final project grade obtained by the student (collected from the instructor's grades book) was included as a response variable.

The second step in the data analysis pipeline consists of loading available instances into the Ranker plug-in. Once this step is performed, all students are available for

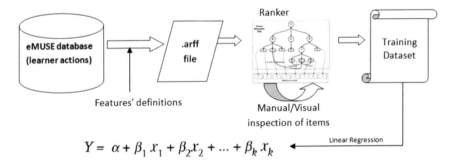

Fig. 1 General structure of the data analysis pipeline

inspection in a ranked list. An excerpt from such a ranked list is included in Fig. 2, with students from the first year of study.

The ranker tool visualizes the ranked list and implements functionalities for marking and deleting instances. These operations are intended to be performed by the data analyst during the analysis process. The first practical usage of the tool regarded semi-automatic removal of two categories of students: ones with considerable social activity and low grades and ones with low social activity and high grades. We observed that the first category of students artificially made a lot of activity on social media tools to obtain higher grades. The most common actions were to post many irrelevant messages on Twitter, revise wiki pages without changing the content, etc., and all these activities may be considered spam. The hypothesis is that we may obtain a more realistic, better, relevant, and interpretable model by eliminating these instances. Intuitively, we assume that such instances (i.e., high activity in social media, but low grades and low activity in social media, but high grades) do not represent the students that we want to model in our data analysis and should be discussed separately and from other perspectives.

We have used the linear regression algorithm in Weka for the data analysis itself as the first method. The goal of this study was to determine the impact of each feature by computing its coefficient in the linear expression. Since cohorts of students may be distinct from one year to the other, we decided to create a leaner model for each year of study.

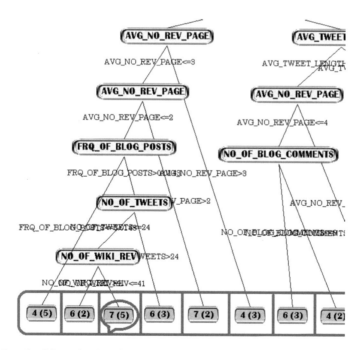

Fig. 2 Sample of the ranked list of students

The proposed approach for building an accurate data model for identifying and deleting the *"spam"* and *"don't care"* users and obtaining the input dataset that is used for further processing [18]. Thus, the input data set consists of instances (i.e., learners) defined by three sets of features, one set for each social tool (i.e., Twitter, blog, and Wiki) and one target variable represented by the exam grade.

Figure 3 presents the data processing pipeline. The baseline model [19] is further used to check the impact on data and classify learners after computing and scaling the principal components (PC) values. The PCs based classifier is compared with the baseline classifier in terms of their ability to classify learners correctly.

Principal Component Analysis (PCA) [20, 21] is used as a dimensionality reduction technique. That is why we need to ensure that the newly computed features can also build a high-quality classifier along the data processing pipeline. This approach is regarded as a *"sanity check"* such that in the situation in which the PCs based model has a high decrease in terms of accuracy, it would be a clear indication that processing of the initial input dataset has degraded the data quality in a large extent and thus it cannot be further used.

Computing the PCs is performed in two ways. One approach computes the PCs at the tool level, reducing the number of features from the available ones at the tool level to the most significant one. The obtained PCs, for a tool, is further regarded as a digest of input features.

After obtaining the PCs from input features, we choose the most representative one, such that it captures a high amount of variance in the input data. Figure 4 presents the general approach of PCA procedure that transforms the original raw *"k"* features from a tool (i.e., Twitter) to k orthogonal (i.e., independent) features. For example, suppose the PC1 (the most representative component from k orthogonal computed

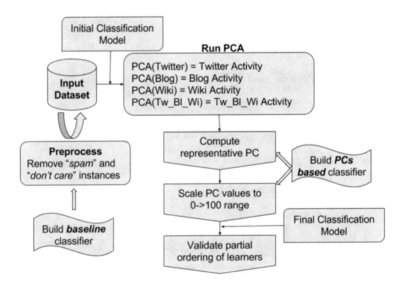

Fig. 3 Overview of the data processing pipeline

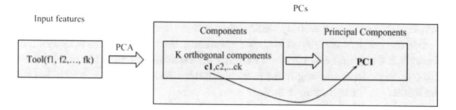

Fig. 4 Dimensionality reduction for each tool

components) captures over 90% of the variance from computed PCs. In that case, it means that we may reduce the number of initial features from k down to one with a minimum loss of information.

From this perspective, we may be in the situation when summing over 90% needs more than one feature, and that is why we need more PCs to represent the input data. As a goal of our data analysis process, we aim to have only one PC representative for each tool or even for all social apps. This approach has the advantage of obtaining a single reduced value for each instance instead of k first input features.

As the percentage of the PC is higher, we are more confident that the obtained value is better representing the initial input data. The main disadvantage of this technique is the poor interpretability of the computed PC value. In conclusion, this step has the advantage of computing a minimal number of features representing the data and the disadvantage of losing interpretability from the first features.

Another helpful interpretation of results that may be performed at this step regards identifying the first attributes that have the most significant impact on obtained PC.

Coping with the lack of interpretability on obtained PC values has a significant negative impact on further analysis. For example, given two instances with initial input feature values, a domain expert may visually compare in an empirical way the activity values at the feature level and interpret them in such a way that the most active learner can be assessed. From this perspective, obtained PC values for the identical two learners cannot be compared such that we cannot say which one had more activity. Another critical issue is that we cannot interpret the PC values for indicating whether one learner or another had a small or large activity level on a particular tool.

The proposed solution for tackling the problems of poor interpretability is to scale PC values to [0, 100] range in such a way that 0 means "*no activity*" and 100 is interpreted as "*large activity*."

Usage of this approach orders learners in terms of their newly computed PC value and keep the original input features' values in relation to its initial PC value. We perform this transformation in two ways: visual and formal. The visual analytics picks several instances with low, average and high values in the initial PC value and comparatively analyses original input features. Keeping in mind that a domain expert has a clear intuition on the meaning of the values of the original input features, the partial ordering of learners in terms of their activity level can be performed visually.

We also prove by formal analysis of the entire dataset of learners' correlation between the performed activity values and corresponding scaled PC values.

For the original activity values, we consider a learner with zero activity on all features as the origin and compute the Euclidian distance to all learners. Naturally, if the distance from the origin to a learner is more considerable is an indication that the learner performed more activity.

Definition: We define the "*origin learner*" as the learner with all original feature values set to zero, alias a virtual learner who has not performed any action on any tool.

Definition: We define the "*activity distance*" to be the Euclidean distance from the "*origin learner*" to an *actual* learner.

$$ED = \sqrt{\text{twitter}^2 + \text{blogspot}^2 + \text{mediawiki}^2}$$

We compute the Euclidean distance from origin to the learner by using as the origin (0, 0, 0) coordinates all three computed activity values for Twitter, Blogspot, and MediaWiki.

$$\mathbf{Di} = \text{Euclidean Distance ("origin learner''}, \text{Learner}_i)$$

This definition allows us to compare the performed activities of learners. Intuitively, if learner A is further from "*origin learner*" than a learner B, it means that learner A performed more activity than learner B.

This approach allows us to obtain the value of the distance from "*origin learner*" for each learner. We formally evaluate the correlation between distances from "*origin learner*" and scaled PC values.

From this perspective, evaluating the correlation in terms of confidence interval validates the ranking obtained by the scaled PC values by assessing its quality in terms of partially ordering learners according to their performed activities. The worst-case scenario may occur when a large residual error is obtained. Thus there is no correlation between initial raw data and obtained PC values which can be explained by a massive information loss in the dimensionality reduction process. A positive correlation indicates that original data may have been reduced with a minimal loss of information. A negative correlation would also be not satisfactory for the data transformation since this would mean that an increase in actually performed activity is regarded as a decrease in distance.

Finally, from a classification perspective, we verify that the final obtained classification model has similar accuracy as the initial one. This validation makes sure that all performed data analysis steps have not significantly altered the initial model and thus can be further used with the same confidence. From a data analytics perspective, the impact of running PCA and scaling is highly dependent on initial data quality. Running PCA should keep only one or a maximum of two PCs that highly represent the initial data while scaling should make the values highly interpretable. The final two steps—correlation of distances with scaled PC values and rebuilding for evaluation of the initial classification model—provide sound evidence on the quality and interpretability of the obtained scaled PC values and associated classification model. At this step, the focus is not on acquiring a highly accurate model but on decreasing accuracy. Thus, a high decrease of precision and a low correlation indicate that initial processing steps (i.e., PCA and scaling) have altered the data too much, and the resulted features cannot be further used for classification reliably. Further processing may become necessary to improve the baseline classification model regarding feature selection, outlier/noise reduction, and proper fitting of the used classification algorithms.

The overall data analysis procedure uses interpretable scaled computed PC values (at tool level or general) with a high correlation with actually performed activities. Intuitively, a worst-case scenario may occur when scaled PC values represent in an unfortunate way the input data and thus exhibit a low correlation with a small confidence interval. The scaled PC values cannot describe the performed activity as a digest in this regrettable situation since the partial ordering is not valid. Therefore other dimensionality reduction techniques need to be used.

2.2 Finding the Learners that Simulate Activity and Explore the Correlation Between Social Activity and Learning Performance

WEKA is one of the most widely used open-source machine learning libraries that offers a tool with a graphical interface and a *.jar* file that can be included in a Java project. This library encapsulates algorithms from several classes like supervised and unsupervised learning or association rules. There are many classifications, regression and neural networks algorithms in the section where supervised algorithms are implemented. There are also various classification approaches starting from decision trees, rule-based algorithms, and boosting or offering the user the possibility of building bagging or voting ensemble algorithms.

Decision trees (DT) are a part of classification techniques that use trees as baseline models. The model is composed of nodes, leaves, and vertices, and for the algorithm used in this chapter, the nodes represent decisions, and the leaves hold the instances. For our approach, we use J48 which implements the C4.5 algorithm in Weka. There are many approaches to building a decision tree, but for J48 every node represents a split that holds a decision, and every leaf has a rule that leads to it. More precisely, starting from the root and going to a leaf by parsing a set of nodes, we get a rule, and in the leaf, we have the instances that obey that rule. One key aspect that needs to be mentioned is that many rules may lead to the same class or label in a decision tree. Different leaves may be in the same class even if the instances don't share very close properties due to non-linear characteristics, the decision boundary.

For preprocessing, we use "*InfoGainAttributeEval*" for feature selection tasks. This algorithm mainly measures how each feature contributes to decreasing the overall entropy. A useful attribute is an attribute that contains the most information, so it will profoundly reduce overall entropy. For the learning algorithm, it is essential to find this attribute and then assign it a weight.

$$\textbf{InfoGainAttributeEval}\ (\text{Class},\ F1) \ = \ H(\text{Class}) \ - \ H(\text{Class}|F1)$$

To compute $H(\text{Class} \mid F1)$, we need to split the dataset; each branch has its entropy, and we need to compute the entropy for each split. $H(\text{Class} \mid F1)$ then equals the sum of both children's entropy, weighted by the proportion of instances that were taken from the parent dataset.

The primary goal of the data analysis pipeline is to detect "*spam*" and "*don't care*" users. We define a "spam" (S) user as a learner with a high activity level and low final grade. The intuition for an S user in an online educational context is given by the attempt to obtain a higher grade by spamming activities. More clearly, S users do not explain their low final grade by the significant social media activity level. We define a "*don't care*" (DC) user as a learner with low activity level and high final grade. The intuition for a DC user in an online educational context is given by the high final grade that has been obtained and is not explained by the low social media activity level. Thus, the significant value of the final grade may be defined by other indicators that do not regard the activity on the social platforms (e.g., additional learning activities, for which the instructor may provide bonus points).

The data processing pipeline consists of several modules and sub-modules that perform feature selection, model building, labelling and ranking of users.

2.3 Engagement by Alerts

We have developed one practical application of generating alerts [22] based on the activity data available in the eMUSE database. The main characteristics of the data-driven analysis are presented in Fig. 5.

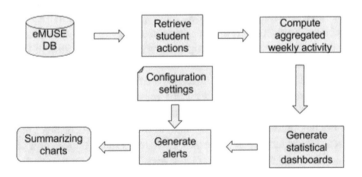

Fig. 5 Data analysis workflow

The developed system determines the weekly activity levels performed by students in all platforms and of all types, such that obtained dataset may be further used in the data analysis process and computing statistical dashboards along with t-tests and p-values. The analysis process is parametrized by window size and overlap, thus fine-tuning the frequency with which students receive alerts.

3 Experiments and Results

3.1 Gathering Data from Several Social Media Platforms

The total count of raw student actions recorded by eMUSE was 6030 in first year and 3604 in second year. There were four students that had no activity on any social media tools (3 in Year 1 and 1 in Year 2), so they were omitted from the analysis. Thus, the total number of students studied further is 63 in Year 1 and 52 in Year 2. Starting from the raw data, the values of the 8 attributes introduced in the previous section are computed.

Two .*arff* files, related to the datasets from Year 1 and Year 2, are consequently generated. A sample from the *arff* file for Year 1 is presented below:

```
@DATA
...
5,  0.0571,  0,  0,    0.00,  9,   2, 2, 4
13, 0.1714, 40, 69, 0.5714, 76, 26, 2, 8
```

Each row in the dataset represents the computed attribute values for a student. For example, the recorded social media activity for the first student consists of 13 blog comments, a frequency of 0.1714 blog posts per day, 40 tweets, an average tweet length of 69 characters, a frequency of 0.5714 tweets per day, 76 wiki page revisions, 26 wiki files uploads and an average of 2 wiki revisions per web page. The student obtained the final grade 8; a 10-point scale grading system is used, 5 being the minimum grade for passing. By consequently visualising the data with our Ranker plug-in, we identified students that are not convincing for the research and thus induce a bias in the data model, as described in the previous section.

The next step was to run the simple linear regression algorithm from Weka on the remaining instances included into the datasets; the obtained coefficients are in Table 1, and the quality metrics for the derived models are in Table 2 (we used the percentage split method for validation).

As it can be seen, the correlation coefficients are over 0.6 in both years, indicating a relatively strong relationship between the final student grade and the corresponding regressors. In particular, the grade is influenced by the frequency of blog posts and the average length of the student's tweets in both years. However, the other two grade predictors are different from one year to another: (i) the number of wiki page

Table 1 Coefficients for linear regression models

	βs	Year	
		Year 1	Year 2
Intercept	β_0	3.757	4.521
Blog-related features	β_1	0	0.277
	β_2	2.751	2.639
Twitter-related features	β_3	0	0
	β_4	0.032	0.027
	β_5	0	0
Wiki-related features	β_6	0.021	0
	β_7	0	0.091
	β_8	0.293	0

Table 2 Quality metrics of the derived models

Quality metric	Year	
	Year 1	Year 2
Correlation	0.628	0.760
Mean absolute error	1.453	1.322
R^2	0.542	0.517

revisions and the average number of revisions for each distinct wiki page in Year 1; (ii) the number of blog comments and the number of files uploaded on the Wiki in Year 2.

We further analyse the level of impact of each regressor by computing the value for which the final grade is increased by one point, given that all other attributes values remain unchanged. Therefore, in first year, a one-point difference in the mark would roughly correspond to either: (i) 0.364 more posts per day (i.e., about 25 more blog posts per semester); or (ii) an average tweet length increased by 31 characters, or (iii) 48 more wiki page revisions; or (iv) an average number of revisions for each wiki page increased by 3. Similarly, in Year 2, a one-point difference in the grade would roughly correspond to either: (i) 4 more blog comments; or (ii) 0.379 more posts per day (i.e., about 27 more blog posts per semester); or (iii) an average tweet length increased by 37 characters; or (iv) 11 more files uploaded to the Wiki. This data should be interpreted in the context of the average recorded values of the features.

As far as the average absolute error is concerned, both models have satisfactory values (less than 1.5); however, in order to be used in a predictive context, even lower values are desirable. The R^2 values greater than 0.5 indicate that the considered attributes can explain about 50% of the variance in student performance.

Overall, these models show that students who contributed actively on Wiki, blog, and Twitter generally obtained higher final grades, which is in line with previous studies. The slightly different patterns of attributes impacts over the years could be explained by the specificities of each year's learning scenario.

An extension the study on seven years with more interpretable results on student's activity has been performed on 367 instances (i.e., learners) that used Twitter, Blogspot, and Wiki social media tools as learning environments at Web Application Design bachelor course [18].

The first processing step regards choosing the principal component for each subset of the feature. All the features are numeric and represent counts of performed activities, i.e., NO_WIKI_REV—the number of wiki page revisions, NO_WIKI_FILES—the number of files uploaded on the Wiki.

Figure 6 presents the proportion of variance computed for each component. On the horizontal axis, we have all the principal components computed for each tool (the number of components is equal with the number of dimensions), and on the vertical axis has presented the proportion of variance value. Thus, in Fig. 6a, while in for other tools, we have five and nine components, respectively. The maximum number of components is twelve and appears in the case of the Blogspot tool Twitter has only five, and in the case of MediaWiki, we have nine. In Fig. 6a, the line represents the value of the proportion of variance for the Blogspot variable; most of the variance (0.9879) is represented by the first component, and we don't need to consider another one. In the case of Twitter Fig. 6b the first component shows more than half of the variance, an increasing proportion between the first two components represents 0.9859 of the overall proportion of variance. In the last case, the first component represents the MediaWiki tool with 0.9454 of the variances. Therefore, in one practical scenario, the principal component is represented only by the first component, which is further used for obtaining an interpretation of the PCA reduction process. This analysis aims to get the PCs that are further used in the data analysis process with a minimum information loss from original raw data.

As the learners are ordered in non-decreasing order of scaled all features column, we expect that the values in the last column (i.e., distance from "origin learner") to be also in non-decreasing order. Still, as we observe in the results, we may face the situation in which we may have a decrease in distance for the same—or even increasing—activity level. This situation may be observed clearly in Fig. 7 where, for the same—or slight increase—activity value, we may obtain a slight increase in the distance, with a clear correlation. Ideally, if such a situation would not occur, it would infer a high quality fitted regression line with a close to zero error. The worst-case scenario occurs when the fitted line exhibits a significant error with a minimal confidence interval. Thus, in many situations, a significant increase in activity level is accompanied by a large decrease in distance from "origin learner".

Figure 7 presents the computed activity trend based on the learners' activity computed for all features and the distance from the origin. We have the overall learner activity on the OX axis, and on the OY, we have the computed Euclidean distance from origin using the three values computed for each tool. Each blue dot is represented by an instance (i.e., student), and we can see in the chart how close the trend line is. Analyzing the chart, we can see that all the dots (instances) stay very close to the trend line and also follows the trend. Most of them are close to the origin, and as we go further, they become rarer and a bit more sparse because as they produce more activity, they are plotted further from the origin. The value

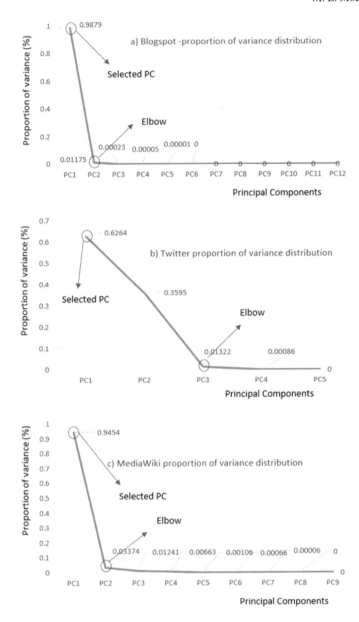

Fig. 6 The proportion of variance distribution for each PC

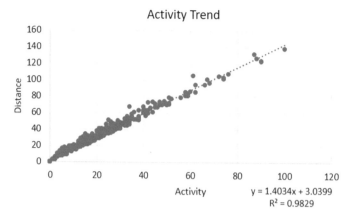

Fig. 7 The activity versus distance to "origin learner" trend

for the intercept is 3.03, showing that the error for "origin learner" is small training into consideration the range values for a distance from 0 to 140. The high positive correlation is also suggested by the significant value of R^2, which is close to one.

This chart offers a visual analytics solution to validate our approach. Data exhibiting a higher variance (i.e., items that are not close to the trend line) would indicate the lack of correlation. A significant intercept value would be interpreted as predicting a considerable distance from "origin learner" for a learner that has not performed any activity, which is misleading in the data analysis process. On the other hand, a lower value in R^2 would be interpreted as a clear indication that the fitted line exhibits a low correlation.

Evaluating the impact of PCA reduction process regarding classification accuracy is performed in a comparative analysis. We have used a set of seven algorithms (i.e., Random forest, j48, c5.0, ctree, evtree, JRip, OneR and kk-NN). Each used algorithm computed the classification accuracy for a trained model on initial feature values (before PCA reduction), on PCs obtained for each tool, and on one PC obtained from all available features. The dependent variable is represented by the final grade, which is nominal and has three values: low, average, and high. The results show that accuracy is the same (i.e., 75%) for *ctree* algorithm, but it slightly decreases in most cases. In one situation (i.e., "oneR" algorithm) the accuracy slightly increases. Thus, we conclude that PCA reduction process has no significant impact on classification accuracy and thus, from this point of view, the compressed data profoundly represents the original data. Although the obtained classifiers may not be used in a practical context due to their low accuracy, the determination of this analysis is aimed to get a clear indication regarding the decrease in data quality, not to train a high-quality classifier.

3.2 Marking the Learners that Simulate Activity and Analyze the Correlation Between Social Activity and Learning Performance

3.2.1 The Dataset Description

The dataset used as input comprehends all the activity traces gathered after five years of study [23]. The total number of instances is 285: 39 in Year 1, 56 in Year 2, 63 in Year 3, 52 in Year 4 and 75 in Year 5. Thirteen learners with no activity were not taken into consideration in the input dataset. From the classification task perspective, they are instances with identical zero valued features and distinct class labels and therefore no decision boundary may be obtained for them.

3.3 Exploring the Impact of Social Media on Students' Performance

We defined and computed a number of 28 features that describe an instance of the input dataset that is a learner. We further present the feature definitions from the .arff file. The following Table 3 describes all features that were defined and computed. Table 4 presents a fragment from the aggregated activity counts of five students from year 2015/2016. The first column represents the student's ID, and the next nine columns include the activity performed in weeks 4–12.

For example, if we consider the student with ID 6, for a window that has four weeks and no overlap (e.g., first window composed of weeks 4, 5, 6, 7 and second window composed of weeks 8, 9, 10, 11) we have a downward activity trend; if we have the same window size but with an overlap of two (e.g., first window composed of weeks 4, 5, 6, 7 and second window composed of weeks 6, 7, 8, 9) we may not be able to say for sure what type of trend we have. In this case, applying t-test can help identify students' activity trends and corresponding states.

In order to investigate the number of prospective alerts that would be issued, we ran the analysis with different values for the configuration parameters (window size, overlap, and p-value). More exactly, we used three values for p (0.25, 0.5 and 0.75), three values for window size (2, 3, 4) and four values for overlap (0, 1, 2 and 3). We can thus explore the impact of all configurations on the computed number of alerts, as presented in Fig. 8.

By analysing the chart in which w means the window size and o means the window overlap, we notice a pattern in the prospected average number of alerts as a function of available configurations. This leads to the assumption that we may confidently choose a particular configuration of parameters for subsequent academic years and estimate the average number of alerts that will be issued. It may also be noted that the same number of alerts can prospect for two or more distinct configurations. Therefore the instructor may choose the one that better fits the particular course settings.

Table 3 Features description

Feature name	Description
FRQ_OF_BLOG_POSTS	The frequency of blog posts
NO_OF_BLOG_COM	The number of comments posted on the blog
NO_BLOG_POSTS	The number of blog posts
AVG_RESPONSE_TIME_BLOG_COM	Average response time for a blog comment
AVG_BLOG_POST_LENGTH	The average length of a post made by a student
AVG_BLOG_COM_LENGTH	The average length of a comment made by a student
NO_ACTIVE_DAYS_BLOG	The number of days in which a student was active on the blog
NO_ACTIVE_DAYS_BLOG_ POST	The number of days in which a student was active on the blog
NO_ACTIVE_DAYS_BLOG_ COM	The number of days in which a student was active on the blog
NO_ACTIVE_WEEKS_BLOG	The number of weeks in which a student was active on the blog
NO_ACTIVE_WEEKS_BLOG_ POST	The number of weeks in which a student was active on the blog
NO_ACTIVE_WEEKS_BLOG_ COM	The number of weeks in which a student was active on the blog
AVG_TWEET_LENGTH	The average length of the student's tweets
FRQ_OF_TWEETS	The frequency of posted tweets, measured in tweets per day
NO_TWEETS	The number of tweets
NO_ACTIVE_DAYS_TWITTER	The number of days in which a student was active on Twitter
NO_ACTIVE_WEEKS_ TWITTER	The number of weeks in which a student was active on Twitter
NO_OF_WIKI_REV	The number of wiki page revisions
AVG_NO_REV_PAGE	The average number of revisions per each distinct wiki page
NO_ACTIVE_DAYS_WIKI	The number of days in which a student was active on the Wiki
NO_ACTIVE_DAYS_WIKI_ REV	The number of days when a student was active on the Wiki
NO_ACTIVE_WEEKS_WIKI	The number of weeks in which a student was active on the Wiki
NO_ACTIVE_WEEKS_WIKI_ REV	The number of weeks in which a student was active on the Wiki
NO_OF_WIKI_FILES	Number of files uploaded on the Wiki
NO_ACTIVE_DAYS_WIKI_ FILES	The number of days when a student was active on the Wiki

(continued)

Table 3 (continued)

Feature name	Description
NO_ACTIVE_WEEKS_WIKI_ FILES	The number of weeks when a student was active on the Wiki

Table 4 Sample activity dataset

Student ID	W4	W5	W6	W7	W8	W9	W10	W11	W12
5	1	2	9	0	0	3	1	0	0
6	1	2	27	66	3	2	1	2	0
7	24	4	2	2	4	1	3	1	0
8	46	3	65	18	3	34	7	19	33
9	17	0	27	17	0	3	14	2	0

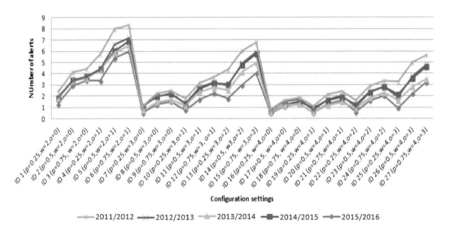

Fig. 8 Average number of alerts per student, for all available configuration settings

For example, suppose the instructor wants to have an average number of three alerts per student in the next academic year. In that case, he/she may choose among several available configurations, as presented in Table 5. If a higher number of alerts is desired (i.e., over 6), then fewer configurations are available.

Let's discuss the impact of several specific configurations on the number of released alerts briefly. As a general rule, we can see that as the value of p rises, the number of alerts also rises. Conversely, the number of alters will decrease if the window size grows. For example, if we take two configurations with the same window size and overlap, e.g., ID 13 ($p = 0.25$, $w = 3$, $o = 2$, avg. no. of alerts $=$ 2.92) and ID 15 ($p = 0.75$, $w = 3$, $o = 2$, avg. no. of alerts $= 5.48$), then we can tell that the number of released alerts almost doubles when p-value increases from 0.25 to 0.75. Similarly, if we take two configurations with the same p-value and overlap,

Table 5 Available configurations for an average number of three alerts per student

Config. ID	p-value	Window size	Window overlap	Avg. no. of alerts
2	0.5	2	0	3.44
11	0.5	3	1	2.51
12	0.75	3	1	3.01
13	0.25	3	2	2.92
24	0.75	4	2	2.66
26	0.5	4	3	3.45

e.g., ID 24 ($p = 0.75$, $w = 4$, $o = 2$, avg. no. of alerts $= 2.66$) and ID 15 ($p = 0.75$, $w = 3$, $o = 2$, avg. no. of alerts $= 5.48$) then we can see that the number of alerts more than doubles when the window size decreases from 4 to 3.

4 Conclusions

Two main types of interactions occur in e-Learning platforms: the first one is *between users* and the second one is *within the e-Learning platform*. The first type of interaction (between users) can be improved by supervised learning techniques for recommending learners tutors or other learners with similar interests. The second type of interaction can be accomplished mainly by adapting the interface to the users' needs and context and by recommending learners different learning resources that may help them to improve their knowledge level or at least those resources would be of their interest.

The popularity and impact of social media activity have been based on the usage of blogs, wikis, microblogging tools, social bookmarking tools or social networking services. As *students' participation and engagement are generally increased by the introduction of social media tools*, it is still important, for a correct assessment, to make a proper evaluation of the contribution of social media for teaching and learning, and to study the learning processes that occur in social media. Regarding this context, we need to know how active participation in social media tools correlates to student performance and successful learning results.

In this chapter, we pointed our attention to the less explored area of student participation in social media-based learning settings. More exactly, we explore students' activity on three social media tools (Wiki, blog and microblogging tool); these were used for communication and collaboration support in a project-based learning scenario. The analytics on social activity traces showed detailed insights on students' behaviour and engagement and their impact on the final learning outcome. We aimed towards the obtaining of an interpretable data model that explains students' final mark using social media interactions. Multiple linear regression algorithm was used to build this student model and explore the effects of various types of student activities on the final learning outcome, and various data analysis methodologies were used to challenge the problem of proving the association between academic performance and educational activities performed on social media tools.

References

1. Kemp, S.: Digital, social mobile worldwide in 2015. In: We are social. Retrieved from Digital, social mobile worldwide in 2015. We are social. http://wearesocial.net/blog/2015/01/digital-social-mobile-worldwide-2015
2. Hamid, S., Waycott, J.: Understanding learners' perceptions of the benefits of online social networking use for teaching and learning. Internet High. Educ. **26**, 1–9 (2015)
3. Mao, J.: Social media for learning: a mixed methods study on high school learners' technology affordances and perspectives. Comput. Hum. Behav. **33**, 213–223 (2014)
4. Popescu, E.: Approaches to designing social media-based learning spaces. In: Proceedings of the BCI '15 (7th Balkan Conference on Informatics), p. 40. ACM Press (2015)
5. Gruzd, A., Haythornthwaite, C.: Learning analytics for the social media age. In: Proceedings of the LAK'14 (Fourth International Conference on Learning Analytics and Knowledge, pp. 254–256 (2014)
6. Morris, L.V., Finnegan, C., Wu, S.S.: Tracking student behavior, persistence, and achievement in online courses. Internet High. Educ. **8**(3), 221–231 (2005)
7. Schuster, R.B.: Participation: the online challenge. In: Aggarwal, A. (ed.) Web-Based Education: learning from Experience, pp. 156–164. Idea Group Publishing, Hershey, Pennsylvania (2003)
8. Michinov, N., Brunot, S.: Procrastination, participation, and performance in online learning environments. Comput. Educ. **56**(1), 243–252 (2011)
9. Giovannella, C., Popescu, E.: A PCA study of student performance indicators in a Web 2.0-based learning environment. In: Proceedings of the ICALT 2013 (14th IEEE International Conference on Advanced Learning Technologies), pp. 33–35 (2013)
10. Beutell, J.A.: Performance indicators in online distance learning courses: a case study of management education. Qual. Assur. Educ. **12**(1), 6–14 (2004)
11. Zhang, D., Zhou, L.: Instructional video in e-learning: assessing the impact of interactive video on learning effectiveness. Inform. Manag. **43**, 15–27 (2006)
12. Lu, J., Yu, C.-S.: Learning style, learning patterns, and learning performance in a WebCT-based MIS course. Inform. Manag. **40**, 497–507 (2003)
13. Junco, R., G.H.: The effect of Twitter on college student engagement and grades. J. Comput. Assist. Learn. **27**(2), 119–132 (2011)
14. Stafford, T., Elgueta, H.: Students' engagement with a collaborative wiki tool predicts enhanced written exam performance. Res. Learn. Technol. **22** (2014)
15. Popescu, P.S., Mihaescu, M.C.: Using ranking and multiple linear regression to explore the impact of social media engagement on student performance. In: Proceedings IEEE Computer Society ICALT 2016, pp. 250–254 (2016)
16. Popescu, E., Cioiu, D.: eMUSE-Integrating Web 2.0 tools in a social learning environment. In: International Conference on Web-Based Learning, pp. 41–5. Springer, Berlin, Heidelberg (2011)
17. Mihăescu, M.C., Popescu, P.Ș, Burdescu, D.D.: J48 list ranker based on advanced classifier decision tree induction. Int. J. Comput. Intell. Stud. **4**(3–4), 313–324 (2015)
18. Mihaescu, M.C.: Generic approach for interpretation of PCA results-use case on learner's activity in social media tools. Adv. Electr. Comput. Eng. **18**(2), 27–34 (2018)
19. Mihaescu M.C., P. P.: Questionnaire analysis for improvement of student's interaction in tesys e-learning platform. Roman. J. Hum.-Comput. Inter. **10**(1), 61–74
20. Wold, S., Esbensen, K., Geladi, P.: Principal component analysis. Chemom. Intell. Lab. Syst. **2**(1–3), 37–52 (1987)

21. Abdi, H., Williams, L.J.: Principal component analysis. Wiley Interdiscip. Rev.: Comput. Stat. **2**(4), 433–459 (2010)
22. Popescu, P.Ş.: Generating alerts for drops in student activity levels in a social learning environment. In: International Conference on Web-Based Learning, pp. 62–71. Springer, Cham (2017)
23. Mihăescu, M.C.: Data analysis on social media traces for detection of "spam" and "don't care" learners. J. Supercomput. **73**(10), 4302–4323 (2017)

Usability Evaluation Roadmap for e-Learning Systems

M. C. Mihăescu⊙, P. S. Popescu, M. L. Mocanu, and C. M. Ionaşcu

Abstract Usability evaluation in e-Learning systems represents one of the final tasks while evaluating and increasing student engagement. This chapter tackles the problem of interface optimisation by analysis of students and professors to questionnaires and surveys. The critical aspects revealed are the factors that influence the interaction, such as the number of hours spent weekly or years spent using the platform. We have found that specific factors highly influence the perceived ease of use. Finally, we describe a data analysis pipeline whose goal is to recommend tutors for students with the same goal of increasing engagement. We conclude that the presented usability evaluation roadmap represents a final data analytics step that should be followed whenever building an e-Learning application.

Keywords Interface optimisation · Questionnaire and survey analysis · Perceived ease of use · Engagement

1 Introduction

Applying HCI in online educational environments is challenging because various users use e-Learning platforms with very diverse knowledge regarding them. Many functionalities need to be easily used and understood by the users because there are critical functionalities like testing procedures or taking exams. A low-quality interface may consume a significant amount of time from the learners when taking tests or an exam, leading to low results.

The usage of HCI in e-Learning platforms comes naturally because of the large and diverse number of users that act on e-Learning platforms. One of the most significant problems in e-Learning platforms is the difference between traditional educational

M. C. Mihăescu (✉) · P. S. Popescu · M. L. Mocanu
Department of Computer Science and Information Technology, University of Craiova, Str. A.I. Cuza, Nr. 13, Craiova, Romania
e-mail: cristian.mihaescu@edu.ucv.ro

C. M. Ionaşcu
Department of Statistics and Economic Informatics, University of Craiova, Str. A.I. Cuza, Nr. 13, Craiova, Romania

environments regarding interaction. This gap can be lowered if we provide enough feedback for the professors and an optimised learning experience for the learners. There are two types of interactions that occur in e-Learning platforms: (1) between users and (2) within the e-Learning platform. The first type of interaction (between users) can be improved by using machine learning techniques by recommending learners tutors or other learners with similar interests. The second type of interaction can be accomplished mainly by adapting the interface to the users' needs and context and by recommending learners different learning resources that may help them to improve their knowledge level or at least those resources would be of their interest.

The primary research approach that we present in this chapter relies on surveys. Using this method on specific groups can reveal serval problems or hints for optimisation in many directions. The research goals vary from adapting interfaces to particular needs and context to understanding how users perceive the e-Learning platforms and finding the factors that influence users' ease of use. Another approach is to create sets of users with similar characteristics and explore the impact on the e-Learning environment. This approach can lead us to a better understanding of the relation between users and e-Learning environments and improve it.

2 Related Work

Early research [1] aimed to identify requirements for a tool to browse tutor–student interaction. A MySQL database has been used for logging tutorial events included a computer, student, start time and end time and mining the data, and there are no machine learning techniques described in it. The data analysis was made by visual browsing, and no advanced analysing techniques were presented. Later [2], presents a survey regarding the application of association rule mining in e-Learning environments. The mining process consists of a workflow of 4 steps: collecting the data represented by all the activities that students perform, data pre-processing (i.e., data integration, user identification, transaction identification, and data transformation), applying the mining algorithms (i.e., association rule mining algorithms), and effectively putting the results into use.

This chapter aims to present several practical approaches for finding usability problems and understanding users perceive the application. There are two types of usability evaluations: using surveys and using heuristics.

A high-level interpretation of the Tesys e-Learning environment's interface has been performed in [3]. A small number of responders composed the responder's group size, and the questions were typically about particular functionalities. In this case, the study revealed some slight problems and facilitated us to understand better how easy it is to use the e-Learning environment for both professors and students. In the same line of works [4], investigated one of the Technology Acceptance Model (TAM) [5] components which are perceived ease of use, and we performed the study on a slightly more significant number of responders. Finally [6], explores the perceived ease of use component, but for the professor's interface and the audience of the study were professors who were performing their activity on two distance learning programs and were already familiar with Tesys.

In 2006 [7], presented a new methodology designed for evaluating e-Learning platforms. The authors specify that there are specific features for capturing the attributes of this kind of Web applications. In [8] also tackle the problem of the usability evaluation of the online educational environments. The authors state that there isn't a combined evaluation procedure for e-Learning environments.

Other related research [9] investigates behavioural intention, perceived satisfaction and effectiveness of online educational environments in a study with 424 university students. The obtained results revealed that the user's perceived self-efficacy is a critical factor that significantly influences the satisfaction of using the e-Learning platform.

Other works, such as [10], describe several other usability methods for evaluation of virtual environments, which define and classify several kinds of assessments along with encountered issues. The authors state some specific characteristics of online environment issues like evaluator issues, physical environments issues, issues related to the type of evaluation, user issues and other occurring issues.

One more recent paper [11] combined the TAM with the didactic cycle and found that they are usually underutilised despite the advantages of e-Learning environments. The authors tried to identify what are the factors that potentially influence students toward attitude using e-Learning platforms.

From a usability evaluation perspective, Nielsen [12] has defined the usability problems characteristics of the user interface and experience, which may generate problems for the end-user. The usability inspection uses two measures: quantitative (which counts the number of usability issues in each category) and qualitative (a detailed description of specific usability problems).

The importance of building highly effective interfaces that provide student engagement has been addressed in [13], which shows that students with forum discussions have a higher chance of finishing a course. This approach implies MOOCs (Massive Online Open Courses) and uses machine learning techniques to extrapolate a small set of annotations to the whole forum. These annotations help in two main ways: summarise the forum's state and allow researchers to deeper understand how the forum is implied in the learning process.

In the same line of research [14], predicts the students' performance based on online discussion forums, which are considered as communities of people learning from each other. The key idea is that forums inform the students about their peers' doubts or problems and represent a way to notify the professors about the learner's knowledge.

3 Proposed Approaches for Interface Optimisation

3.1 Interface Optimisation by Usability Analysis

Usability evaluation of Tesys e-Learning platform is a case study of task-based usability inspection that resembles the heuristic walkthrough. We defined 14 usability heuristics that are further used to describe and document each usability issue and train evaluators. The heuristics regard **user guidance** (i.e., *prompting, feedback, information architecture* and *grouping/distinction*), **user effort** (i.e., *consistency, cognitive workload,* and *minimal action*), **user control and freedom** (i.e., *explicit user actions, user control, flexibility*) and **user support** (i.e., *compatibility with the user, task guidance and support, error management, help and documentation*).

The evaluation implied four experts that tested the platform using a task-based approach with the determination to anticipate the problems of an end-user. Before starting the evaluation process, the evaluators acknowledged the set of usability heuristics and the evaluation tasks. The evaluation tasks on which have been performed are of two types: **teachers' tasks** (i.e., *adding homework, adding a new course/document, sending a message to students* and *verifying a student homework*) and **students' tasks** (i.e., *changing personal data in the user profile, downloading a course, sending a message to a teacher* and *performing a test*).

The evaluation has been performed in two different steps: **individual evaluation phase** (i.e., *each evaluator tested the platform independently and logged the usability issues for each job*) and (i.e., *removing the duplicates, agreeing on a list of unique usability issues, removing the false usability issues, deciding on the severity of the problems, and finalising the issue description*).

The consolidation per job of each usability problem was founded on the "similar changes" principle. For each usability problem, the following information was recorded: anticipated difficulties, situation, context, cause, several suggestions for fixing, usability heuristic violated and severity.

We present the interface optimisation studies for Tesys e-Learning platform [15]. The studies explore how well interface and functionalities were designed in terms of usability. The goal is to gather detailed knowledge about usability issues such that solutions may be further derived. The studies were conducted using general and more detailed surveys by exploring some of TAM components on different audiences, from users who never saw the platform before to users who were using the platform for several years. Users are students and professors as their interfaces differ, and we needed to optimise the whole e-Learning platform, not only part of it. First studies were done using a custom-made survey intended to understand how optimised the platform is. Still, later we moved to more standardised ones, trying to understand how the users perceive the platform by exploring TAM.

The first study is general, and its results show there are no severe issues as most of the audience gave pretty good feedback.

The study uses three surveys that aim to validate the communication section, the testing activity of students, and the interface available for the professor and student.

The communication module covers a significant part of the interaction between students, professors and secretaries that perform their activity within Tesys e-Learning system. The evaluation of the communication section is achieved using eleven mixed type questions. The survey aims to find where we need to improve the interface and the functionalities of the communication section. The first questions evaluate if there are issues with buttons and controls dimension or placement, and the rest of the survey considers the main functionalities of this platform's section.

The testing module evaluation survey has twelve mixed questions that address both the module's interface and functionalities. In the first step, we aim to evaluate if the questions from the testing procedure are well presented, and in the second step, we assess the main controls. Both are significant because we don't want students to waste their time finding a button or a checkbox or having difficulties in reading the question. Other critical aspects refer to the formula used for computing grades and how often students got a question repeated as for each test, the set of questions is randomly chosen.

The general interface evaluation section of the survey has fourteen questions divided into seven for the professor's interface and seven for the student interface. There are two interfaces for both students and professors that are evaluated in this survey. Each question from the student's interface and most from the professor's interfaces will get a mark between 1 and 5 regarding a specific control (button or checkbox) or functionality. In this survey, we explore the interface design to provide the best usability results in the next iteration.

For the experiment's setup, we used a group of twelve learners from different years of study. There were no professors or tutors involved in this research. The motivation for choosing these students regards their academic results and a significant level of trust. The selected students started to use the platform for the first time, so there was no previous experience, and we tried to evaluate how well they adapted to the e-Learning platform. The group of selected students learn at different years of study. This setup was chosen because students also use the e-Learning platform from several years of study from various faculties.

The approach for analysing critical issues that influence the interaction in e-Learning platforms [16] is based on a survey in the Romanian language to which students from the Faculty of Economics and Business Administration (FEBA) and Faculty of Letters (FL) from the University of Craiova (UCV) have answered. The motivation for choosing them is that they used Tesys e-Learning platform for different periods: one, two, or three years. We received 114 surveys from FEBA and 31 from the FL. Regarding surveys distribution, from FEBA we received 42(36.8%) from the first year of study, 35(30.7%) from the second year and 37(32.5%) from the last year of studies; from the FL, we received 18(58.1%) from the first year, 10(32.3%) from the second year of studies and only 3(9.7%) from the last year of studies. The distribution over the years reveals that more students from the first year were interested in participating in such a study. Before analysing the surveys, we eliminated five from FEBA and one from the FL because of incomplete data.

Nineteen questions were used for this study even if in the survey were more because we had four questions regarding their year of study, their learning program, the time spent on the platform and if they use any other e-Learning platform and more questions with text areas to get feedback and recommendations for several functionalities. The question regarding if the students had used before other e-Learning platforms aimed to see if previous experiences in e-Learning have an impact on the usage of Tesys e-Learning platform. Still, all the answers say that they never used another platform.

The survey with items presented in Table 1 contains three parts, aiming to evaluate one subject: main interface—M1 to M6, testing module—T1 to T5 and the communication module, which includes questions C1 to C9. Each question was evaluated using a Likert scale from one to five, where five express a strongly positive reaction

Table 1 Survey's items for students

Item	Question
M1	How do you evaluate the buttons arrangement on the main page?
M2	How frequently do you use the "Home" button to return home from another page?
M3	How fast did you find chapters in the section "Courses"?
M4	How fast did you find the "Videoconference" button?
M5	How fast did you find the "Auto testing" button?
M6	How fast did you find the homework?
T1	Do you consider the auto testing procedure based on chapter separation to be an efficient one?
T2	While you do the autotest, the question's text was shown good enough?
T3	Regarding the questions, the main controls (e.g., checkboxes, buttons, etc.) were easily accessible?
T4	How fast did you find the marks for correct answers and the answering button during auto testing?
T5	Do you consider that the checkboxes should be more prominent?
C1	How hast did you find the messages for professors/secretaries/colleagues?
C2	Being on the main page, how fast did you find the button "Communication"?
C3	Entering the section "Communication," how comfortable are you with the layout buttons on the page?
C4	How easy did you found a student in the section "communication with students."
C5	How easy was to find the messages for the conversation with professors/secretaries/colleagues starting from the home page?
C6	In the section "communication with professors" how fast did you found the professors from a specific discipline?
C7	The message's text field is big enough?
C8	Did you found convenient sending a message to one or more professors from a specific discipline?
C9	Do you consider that the messages display from the "communication with secretaries" section is well designed?

Table 2 User's characteristics

Variable	Question
YOS	What is your year of study?
POS	What is your programme of study?
WHT	How many hours do you weekly spend on Tesys?

and one express an adverse response to the question. A more significant number of questions referring to the communication module is motivated by the complexity and by the messages received in the survey as feedback. Table 2, listed below, presents the questions that aim to evaluate the user's characteristics.

One critical aspect of this study is the number of hours spent on the platforms. It may influence the user adaptability to the interface, and the operating speed may also improve.

Regarding the distribution of the time spent on the platform at FEBA, 64% of the students spent less than 10 h on the e-Learning platform, 31.6% of the students spent between 10 and 20 h per week and 4.4% spent between 20 and 40 h per week. No student from FEBA spent more than 40 h per week. In the case of the FL, 67.7% spent less than 10 h per week, 19.4% between 10 and 20 h per week and 6.5% for the following two categories: more than 40 h and between 20 and 40 h per week. In both cases, most of the students spent less than 10 h weekly on the platform and for this study, we used the first group being formed by the students that spent less than 10 h and the second group with more than 10 h per week.

In terms of distribution of years of study attended by students at FEBA, they were almost equally distributed, but in the case of FL, more than 50% of the students were from the first year of study. This per cent can influence the results of the survey because they had less experience with the platform.

The submitted surveys use 5 Likert scales that provide a ranking of answers, not grades for answers. Considering the answers like grades would imply a strong assumption that there is equal "distance" between any two consecutive answers to questions. This approach would allow assessing the value of the answer as a numeric type, and therefore further computation of analysis of variance and mean values. Because the values are ranks, it led us to use nonparametric analysis of variance, and therefore the Wilcoxon test was selected and used to run on data.

The purpose of the Wilcoxon test was to reveal the questions whose p-value is smaller than the 0.05 threshold value, which represents a clear indication that there is a significant change in perceived ease of use among the students' groups differentiated by a particular factor.

Wilcoxon test (sum of signed ranks) is a test statistic used for statistical hypothesis testing. Computation of the signed-rank reduces to computing the difference among the groups, assigning rank, and finally computing the sum of all ranks.

The approach for exploring how professors perceive the ease of use of Tesys e-Learning platform [6] is based on a study in which 34 professors (18 from FL and 16 from FEBA) completed the survey. The survey has twenty-four questions divided

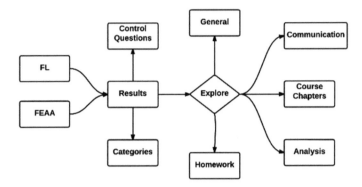

Fig. 1 Study high level design

into three main categories: (1) *EU*, questions for evaluating platform's ease of use (EU), (2) **C**, control questions and (3) **GS**, questions used for group separation.

Figure 1 presents the study's high-level overview. We chose to evaluate the FL and FEBA, from which we gathered the results contained the studied questions. We also added control and categories questions on which we performed the statistical analysis. We explore five principal categories: course module, homework module, the students' analysis module, a communication module, and the overall usability of the main interface.

Regarding exploring the ease (EOS) of use components, we proposed 14 questions meant to reveal specific problems. These EOS questions were meant to be validated by six control questions: four questions for each set of questions, one for the overall survey validation and one for explanations regarding the help section that also validates the previous question because if the response was low, but no suggestions were made there is a certain inconsistency. Suppose there was a contradiction between the control question and the group of questions that were validated by it. The set of answers from that responder was disregarded. Four group separation questions were used, and based on them, we can also get insight into the level of practice of the responder and how often they use the e-Learning platform.

Table 3 presents the questions included in the survey. First, we have the question number; in the second column is the text of the questions and the last column defines the category of the question.

3.2 Experiments for Interface Optimisation for Better Usability

Interface optimisation is a complex problem that needs to be addressed using several techniques because it is dependent on the application audience. We used the survey

Table 3 Questions for professors along with their type

No	The question	Cat
1	How easy is to find the teaching courses?	EU
2	Have you ever used the administration section of a course?	C
3	Organising homework, chapters, and links on the course administration page are intuitive?	EU
4	How simple is to add a new course chapter? Mark from one to five!	EU
5	How simple is to set up a new question for the course chapters? Mark from one to five!	EU
6	Until now, have you ever used the homework management function?	C
7	How simple is to add new homework? Mark from one to five!	EU
8	How simple is to provide grades to homework? Mark from one to five!	EU
9	The functionality for adding external web references (links) for the disciplines is simple to use?	EU
10	How quickly did you find the external web references (links)?	EU
11	Have you ever used the student's analysis module from the disciplines?	C
12	The data provided by the analysis module referring the students was satisfactory enough?	C
13	How bright is the information regarding the students presented in the analysis module?	EU
14	During the past experience with the e-Learning platform, have you ever used the messaging functionality?	C
15	How simple was to use the functionality "communication with students"? Mark from one to five!	EU
16	How simple did you used the section "communication with secretaries"? Mark from one to five!	EU
17	How simple did you used the section "communication with professors"? Mark from one to five!	EU
18	The "Help" section offers adequate data?	EU
19	If you choose the "no" option, please describe what it should contain!	C
20	What is the number of disciplines you teach In Tesys?	GS
21	At what year of study do you teach?	GS
22	How many years of practise do you have at distance learning?	GS
23	How many hours per week do you use the e-Learning environment?	GS
24	What is the overall grade for the platform?	C

method to get an insight regarding the usability level and produce significant recommendations. Firstly, we conducted a survey to understand how students and professors perceive the interface and how easy it is to use it. We explore factors that influence their activity, and in the last surveys, we explored several components of the TAM.

The first experiments conducted to understand Tesys e-Learning Platform's interface were divided into three modules (communication, testing, and general interface) and on both student and professor roles [3]. Below we present the results obtained from every survey along with a short description. For every survey, we completed a table with the results on the columns the questions and on the rows the answers from every student that took part in the study.

For the first questionnaire, we proposed eleven questions; the obtained results are presented in Table 4. Each question received a grade or a letter corresponding to the selected answer. The last three questions that got the same letter: *a,* were referring if the dimension of the field of the message is large enough (question number 9), if the title of the message is clear and big enough (question number 10) and if the whole experience with the platform is good enough. Table 4 aggregates the results obtained after completing the communication survey.

The messaging/communication section includes two types of questions: grade and free choice (multiple answers). Three questions (i.e., 9, 10 and 11) got a single answer. The decision is that the fields dimensions and the user experience regarding the communication section are good enough, and we don't need to implement further modifications.

Based on the answers, we found that the location of the "Communication" button is correct because only 8% of the asked students didn't find the button at first eyesight. Considering question 6, we can conclude that several students had difficulties sending a message to the teacher. After a difficulty analysis over the sixth question, we found some software implementation issues, and we need to test further and solve them.

We have only two questions from the second category that obtained grades above eight for buttons placed on the interface. In conclusion, the controls are well established, and the user's experience with this section is decent. The overall analysis reveals some problems that need to be resolved, but the overall usability is pretty good based on the answers from question no. 11.

Table 4 Results on the communication survey

Uid/qid	1	2	3	4	5	6	7	8	9	10	11
1	NA	a	5	9	9	d	6	b	a	a	a
2	b	b	5	8	9	d	5	b	a	a	a
3	c	a	9	9	9	b	8	c	a	a	a
4	d	a	7	10	10	b	9	a	a	a	a
5	a	a	8	10	10	b	10	c	a	a	a
6	a	a	9	9	10	b	10	a	a	a	a
7	b	a	10	9	9	b	10	a	a	a	a
8	d	a	9	1	10	b	1	b	a	a	a
9	b	a	9	8	10	b	9	b	a	a	a
10	a	a	7	8	7	b	0	a	a	a	a
11	a	a	10	10	10	b	9	b	a	a	a
12	a	a	8	10	8	b	7	a	a	a	a

Table 5 Results for the testing survey

uid/qid	1	2	3	4	5	6	7	8	9	10	11
1	10	a	10	10	3	a	c	5	10	yes	b
2	9	b	10	10	3–5	a	a	8	9	yes	c
3	10	b	9	9	2	a	a	2	10	yes	c
4	10	a	10	10	2	b	c	2	10	no	c
5	8	a	10	8	3	b	a	4	9	yes	a
6	7	b	10	8	3	a	a	2	8	yes	a
7	9	c	10	9	2	b	a	1	9	no	d
8	10	a	10	10	3	b	a	10	10	yes	c
9	8	a	10	10	2	a	c	1	10	yes	c
10	10	b	10	9	10	b	c	1	10	yes	c
11	9	a	10	10	1	b	a	1	9	yes	a
12	8	b	10	10	9	b	a	4	10	yes	b

Table 5 presents the results obtained from the survey used on the testing module. Like in the case of the communication module evaluation, we have the same two sorts of questions for the testing module. For questions 1, 3, 4, and 6, which have multiple match answers, these questions got a mark better than 9, so there is no need for improvement on the tested aspects. Based on question number 8 analysis, we conclude that we need to optimise further the students searching results. Still, a period close to three seconds for searching professors is good enough considering that the functionality was at first glance. One aspect that was also significant was that we needed to adjust the checkboxes dimension, and more precisely, they needed to be enlarged.

Las two questions referred to the possibility to hack the testing procedure. This procedure is crucial for this module, and we wanted to see if any of the students found a method for this. We left them more time to see but no results. There is no question 12 in the table because the results should have been a short description of the procedure.

Question 5 received an assortment of results, and these results may be somehow blurred, but considering the context of the question, we can have a threshold over the time spent on answering a question. A student cannot spend too much time finding the checkbox related to the correct answer, so we certainly need to improve the dimensions. The motivation for taking such a conclusion is that in most cases, he has 30 s for giving the answer and 3 s or more significant time means at least 10% of the total available time.

The last survey presented in Table 6 evaluated the buttons and controls situation for both students and professors. We propose seven questions that refer students, each getting a grade on a scale from 1 to 10. Excepting question number 6, which got a mean grade a bit lower than 6, every question from this set got an average grade

Table 6 Results collected from student's interface survey

Uid/qid	1	2	3	4	5	6	7
1	10	10	10	10	10	6	9
2	10	10	10	10	10	8	10
3	10	10	10	9	10	7	9
4	10	10	10	10	10	1	9
5	10	10	10	10	10	9	10
6	10	10	10	10	10	7	9
7	10	10	10	10	10	8	9
8	10	10	10	10	10	8	9
9	10	10	10	10	10	7	10
10	10	10	10	10	10	5	10
11	4	10	10	10	10	3	9
12	10	10	10	10	10	9	9

very close to 10. Getting a lower grade on question 6 is explained by the significant number of steps the user had to perform to find the homework session.

For teachers responses, the grades were also significant, as presented in Table 7. Still, some of the study participants considered that course and homework need to be revised without stating specific requests as suggested by answers to question 2. For this particular case, we couldn't find any improvements to implement, but we can consider this questionnaire to evaluate the teacher's interface. We can always think about reorganising the courses, links, or homework to see if better results can be achieved, and then we will rerun the evaluation.

Table 7 Results obtained on professor's interface survey

Uid/qid	1	2	3	4	5	6	7
1	a	a	10	10	9	10	10
2	a	a	10	10	9	10	10
3	a	a	10	10	8	10	9
4	a	a	10	10	10	10	10
5	a	b	10	10	10	10	10
6	a	b	10	10	10	10	9
7	a	a	10	10	9	10	10
8	a	a	10	10	10	10	10
9	a	0	10	10	10	10	7
10	a	a	10	10	10	10	10
11	a	a	10	10	10	10	10
12	a	a	10	10	8	9	9

3.3 Experiments Obtained from Analysing Key Issues that Influence the Interaction in e-Learning Platforms

Analysis of key issues that influence the interaction had been performed by running the Wilcoxon test on two groups whose factor of differentiation was the time spent on the platform [4]. Thus, the two groups were characterised by the reported weekly time spent on the platform: one group consisted of students who spent less than 10 h weekly on the e-Learning platform and the second one consisted of students who spent more than 10 h. Thus, the purpose of the study was to investigate how students perceived the ease of use of various functionalities (i.e., general, self-testing, communication) related to the weekly time spent online.

Other indicators that may be computed and bring knowledge that correlates well with results provided by the Wilcoxon test are median and mean values of survey answers, the average number of loggings and the average time spent on the platform. Other factors that may discriminate among groups are the study program or the year of study.

Table 8 presents the results of Wilcoxon's test on survey answers given by FEBA students.

Table 8 Descriptive statistics of Wilcoxon test on FEBA student's answers

Item	<10 online hours per week	>10 online hours per week	W	p-value
M1	4.02	4.43	1758.5	0.0913
M2	2.87	2.70	1415.0	0.6191
M3	4.15	4.70	1926.5	0.0031
M4	4.10	4.29	1617.5	0.4257
M5	4.24	4.56	1722.0	0.1245
M6	3.97	4.58	1923.0	0.0052
T1	4.09	4.19	1581.0	0.5949
T2	4.16	4.46	1785.5	0.0619
T3	4.15	4.26	1551.0	0.7302
T4	4.12	4.26	1567.0	0.6543
T5	3.00	2.90	1439.5	0.7314
C1	4.05	4.19	1609.5	0.4766
C2	4.35	4.41	1495.5	0.9973
C3	4.15	4.31	1610.5	0.4657
C4	3.83	4.07	1663.5	0.2968
C5	4.16	4.26	1574.0	0.6188
C6	4.04	4.24	1638.5	0.3712
C7	4.19	4.09	1467.5	0.8565
C8	4.13	4.26	1537.0	0.7991
C9	4.12	4.19	1583.5	0.5850

The first observed result regards the *p* values smaller than 0.05, obtained for questions M3 and M6 and a close result at question T2. Therefore, students with more than 10 h spent online weekly perceived that they may find much faster the chapters, the homework, and understood the text of the quiz questions in a better way. This interpretation of obtained results is also supported by the large gap in mean values of the two groups for M3, M6, and T2. On the contrary, close values for all other questions is an indication that the 10 h threshold between groups did not make a difference, a fact supported by the close mean values.

Table 9 presents the results of Wilcoxon's test on survey answers given by FL students.

The first observed result regards the *p*-value of the C6 question. This observation is an indication that FL students had problems in communication with professors. Regarding this result, we note that the mean grade in a group with more than 10 h is larger than the mean grade in a group with less than 10 h. This result strengthens the conclusion that communication with professors, especially finding specific professors with which a student needs to communicate, is a possible critical issue in interface design.

Table 9 Descriptive statistics of Wilcoxon test on FL student's answers

Item	<10 on-line hours per week	>10 on-line hours per week	W	p-value
M1	3.95	4.1	114.5	0.6850
M2	2.42	2.9	129.0	0.2962
M3	4.19	3.9	91.0	0.5391
M4	4.04	4.4	117.0	0.5917
M5	3.90	4.0	109.0	0.8741
M6	3.76	3.7	102.0	0.9118
T1	3.47	3.7	107.5	0.9295
T2	3.33	3.5	103.0	0.9477
T3	3.57	4.0	113.0	0.7411
T4	3.47	3.9	115.5	0.6594
T5	3.19	3.9	134.5	0.2026
C1	4.00	3.8	82.0	0.3119
C2	4.00	4.3	110.5	0.8184
C3	3.76	3.7	95.0	0.6764
C4	3.76	3.3	83.5	0.3546
C5	3.90	4.1	102.5	0.9284
C6	4.09	3.3	57.0	0.0345
C7	4.09	3.8	85.0	0.3831
C8	4.00	4.0	92.0	0.5742
C9	3.76	3.6	90.5	0.5314

In general, as a comparison of results between FEBA and FL students, several observations need to be made. Firstly, the mean of grades given by FL students is smaller than the grades assigned by FEBA students. This observation may be explained by the fact that FEBA students are more technical than FL students, and therefore technology has a higher acceptance rate of FEBA students. Another supporting aspect is that the Tesys platform runs at FEBA for longer than at the Faculty of Letters. Therefore, FEBA students may be much more familiar with implemented features. This observation is also supported by many first-year FL students that answered the survey who did not previously use Tesys. On the contrary, the distribution regarding the year of study is well balanced for FEBA students.

The sum of the signed-ranks (W statistic) depends on the number of subjects that answer the survey. The big difference between the number of students that attended the study from FEBA and the FL motivates the big difference regarding the value of W statistic for these two groups.

Let's suppose we have some "n" maximum ranks; in the case of FEBA we got 114 learners that answered the study, so we will have $n \leq 114$ when we compute the sum, and in the case of the FL, only 31. If we calculate the sum using the standard formula ($n(n + 1)/2$), we get a maximum of 3277.5 for FEBA and 450 for the FL. These numbers represent the absolute maximum that can be computed for those two groups in case of maximum difference, and it's relevant only for that group. As the W statistic decreases, the difference between the groups becomes smaller, and their p-value becomes more significant. A significant W statistics value corresponds to a small p-value and a sharp distinction between groups. The maximum possible value may never be achieved as it is always possible to have more than one instance with no difference in the dataset, and that rank will never be computed.

3.4 Results Obtained from Exploring How Professors Perceive the Ease of Use of e-Learning Platforms

We considered two group separation questions to explore the perceived ease of use and investigated whether the tested factors somehow influence the perceived ease of use [6]. We omitted for the study questions twenty and twenty-one because of the lack of information gain, most of the professors that completed the survey were teaching in all of the academic years, and the number of disciplines was from one to three for 94% of the professors from the FL and 100% from FEBA.

Regarding the number of years since the professor started to use the e-Learning platform, we have obtained that 67.6% have used it for more than six years. This finding motivates the small number of hours spent weekly. If they already have excellent experience with the platform, they need to spend many hours managing courses, questions, or performing other tasks.

Table 10 Wilcoxon results regarding the number of years spent on Tesys e-Learning environment

qID	p-value	W-statistic	Mean for <6 years	Mean for >6 years
1	0.585	115	4.739	4.181
3	0.736	117.5	4.260	3.818
4	0.864	122	4.565	4.090
5	0.675	137.5	4.086	3.909
7	0.697	136.5	4.347	4.090
8	0.346	102	4.173	3.454
9	0.625	113.5	4.130	3.545
10	0.431	106	4.043	3.545
13	0.788	119	3.913	3.545
15	0.860	121.5	4.130	3.727
16	0.381	149.5	3.782	3.909
17	0.877	131	3.826	3.636
18	0.741	135.5	3.869	3.818
24	0.798	119.5	4.217	3.818

To have balanced groups, we decided to set a threshold of 6 years, and we merged the two groups from zero to three and from 3 to 6 years, and we obtained one, which is referred to as less than six years in the following table.

Table 10 synthesises the results computed after dividing the information gained in two groups constructed on the number of years the professors who answered performed their activities in the e-Learning environment.

The first column from the table represents the question's number, the second the computed w-statistic, the third column presents the p-value, and the last two columns represent the mean values for each group. Considering the collected data, we define two main groups: teachers with experience greater than 6 *years* of study using the Tesys e-learning environment, and professors who have used the e-Learning platform for less than six years. We compute both w-statistic and p-value to get a more proper interpretation of the difference between these two categories.

Considering the results, we can conclude that the number of years spent on Tesys e-learning environment does not influence how they graded the questions.

Figure 2 presents how many hours the users spend on the platform weekly. From this figure, we can say that professors used the platform relatively lightly because most professors spend less than two hours a week on the e-Learning platform. In the following table, we present the results for two groups; we merged the groups that had less than one hour with the group one that spent one to two hours and then the other two groups obtaining two more similar groups on which we could compute Wilcoxon statistic.

Table 11 presents the results obtained after exploring if the number of hours spent on Tesys offers a significant alteration between groups. We choose, in this case, a threshold of two hours spent in one week on the platform.

Fig. 2 Number of hours
spent on the platform

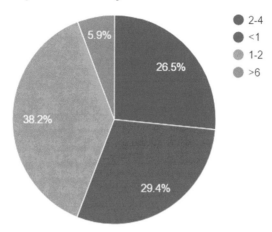

Count of "How many hours do you use the platform weekly?"

- 2-4
- <1
- 1-2
- >6

5.9%

26.5%

38.2%

29.4%

Table 11 Wilcoxon results in the number of hours per week spent on Tesys e-learning environment

qID	P value	W statistic	Mean for <2 h	Mean for >2 h
1	0.961	133.5	4.666	4.5
3	0.588	146.5	4.083	4.136
4	0.136	168	4.333	4.454
5	0.906	128.5	4.083	4
7	0.602	118.5	4.416	4.181
8	0.630	119	4.166	3.818
9	0.632	119	4.166	3.818
10	0.551	116	4.083	3.772
13	0.638	119	3.916	3.727
15	0.744	141	3.916	4.045
16	0.924	129	3.833	3.818
17	0.954	134	3.666	3.818
18	0.362	156.5	3.583	4
24	0.953	130	4.166	4.045

The first column from the table represents the question's number. In the second column, we have the w statistic, and the third column presents the p-value. The last two columns represent the average scores for the chosen categories.

Analysing the results, we can conclude that also, in this case, the number of hours does not influence the perceived ease of use regarding the Tesys e-learning platform.

3.5   Recommending Tutors to Students for Increasing Engagement

One of the approaches for increasing engagement in e-Learning systems may rely upon determining student peers as tutors. Solving this problem reduces to finding students who share similar features and interests and ranking their groups such that tutoring may be enabled.

Since students have available performed activities and their test grades as the target class, we define the task as a classification problem. For example, when using a decision tree as a classifier, the leaves become groups of similar students (i.e., with the same label). If ranking is defined on target classes, they become pools of colleagues that are best candidates for being used as tutors.

The model builder has three main goals: setting up a core data analysis algorithm (i.e., a classifier), a validation technique and a report regarding the error analysis. In this approach, data processing business logic is represented by a classification algorithm implemented by Weka that takes a training dataset a ".arff" file as input. It outputs an in-memory classifier and a serialised XML representation.

The system's validation and error analysis of the algorithm's output enforces us that we have made the right choice. Weka provides measures like accuracy, confusion matrix, kappa statistic, mean absolute error. The recommender module is developed to match the query of a learner and a tutor against the existing data model. From a software engineer perspective, the recommender module is a client for the model builder module. Its main task is to produce results so they may be intuitively displayed by the thin client application used by the learner in his attempt to find a suitable tutor. Therefore, the learner will see a tree-like structure due to the natural shape of the decision trees with the actual class of the learner marked in red and with target classes marked in shades of green. The green classes represent a set of learners suitable for tutors for the learner who is querying the system.

Figure 3 presents how the tutor recommender mechanism is designed as a data workflow. From an interaction point of view, students must query the system that is integrated within Tesys e-Learning platform. After performing necessary operations on the server-side, they obtain the decision tree model filled with prospective tutors. The business logic of the recommender system is built on the server-side. Here the training dataset is created. Afterwards, the data model which is represented as an XML file can be presented on the client-side.

Figure 4 presents how the client–server architecture is designed for the recommender system. When the student requests to view the decision tree, the browser sends the student's ID to the server. The server reads loads *currentStudents.xml* and retrieves features values for currently enrolled students, including those who sent the request. The data collected from the students that have previously used the platform is stored in a file called *repositoryOfStudents.arff*. From a machine learning point of view, this file contains the training dataset that is regularly updated with the information gathered from the current students. Based on the data stored within this file, an XML file called *decisonTree.xml* is generated at the beginning of each semester.

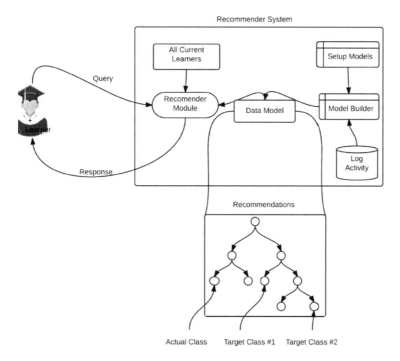

Fig. 3 Activity use case diagram

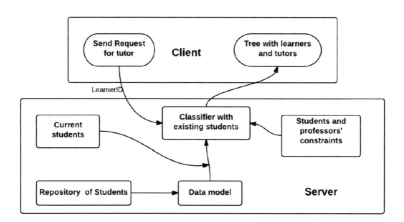

Fig. 4 Data pipeline for the tutor retrieval process

From a machine learning point of view, this file represents the data model against which current learners are classified and tutors are obtained.

The next step in our data pipeline involves using the data model and the query to find the student's actual and target class. Moreover, all the other students are assigned to the corresponding leaf nodes of the tree. Once the process ends, the student can view in the browser the applet with the decision tree, his actual and target class highlighted, and explanations that will assist him in better understanding the structure of the tree. The target node will also provide the list of potential tutors.

The main tasks performed by the recommender module regards building the preliminary data model, indexing currently existing learners into the already created model, setup the recommender system, computing and extracting the relevant tutors by applying the already designed recommendation strategy.

We apply machine learning techniques (i.e., classification by decision trees) to build learners profiles by using already existing implicit performed activities from usage sessions of learners that used the system in previous years. This logged data represents the training data, and the algorithm's output is represented by a set of classes (i.e., leaves of the decision tree) such that each class corresponds to a learner's profile. Once sessions and corresponding activity data are delimited so that all features describing a learner are processed, we can use the decision tree builder to obtain the baseline data model. After the data model is computed, the current students in the e-Learning system are assigned their corresponding leaves. At this step, the existing students are stored in the decision tree structure and are ready to be queried.

The process of finding a tutor for a learner is regarded as a specific implicit query. The parameters aim to tune the finding mechanism such that only optimal results are returned from the pool of possible solutions. The pool of solutions is regarded as a set of leaves containing a group of potential tutors. The collection of classes needs to fulfil one essential requirement: to be labelled with a "better" class label than the one in which the querying learner resides. A total order set of classes is ergonomically computed. The first item contains learners "close" to the querying learner, and the last item containing learners "further" to the querying learner. In this context, parameter tuning will decide a certain number of prospective tutors picked up from one of the target classes. Intuitively, choosing tutors from a class that is "closer" to the querying learners' class will return tutors with a better profile but similar. On the other hand, selecting tutors from a class that is "further" to the querying class will provide the tutors with the best profiles among all students.

The central concept considered in the data model is the "student," which can also be a "Tutor." After the data model is built based on the training data, the current set of students $S = f_{S1, S2, S3...Sng}$ is distributed in similar classes according to the key feature values f_i, k. The features are not weighted for the current prototype implementation since the decision tree provides a ranking in feature selection. All classes of learners are considered as resources for which an "affinity" function needs to be defined to retrieve the most suitable tutors.

Defining the affinity function needs to consider several criteria such as a better overall knowledge weight, specific values in communication-related features (i.e., forum activity, messaging activity, etc.) and demographic characteristics. Due to its particular topology, the decision tree model also ranks the leaves in classes. A normal distribution function is proposed such that the lowest-ranked class is assigned zero-knowledge weight and the highest-rated class is assigned a value of one knowledge weight. All in between classes get a knowledge weight ranking between zero and one. Thus, the data analysis task is to identify the actual class of the learner who is

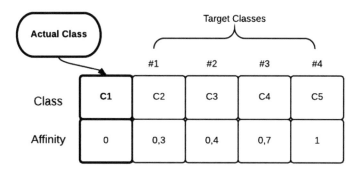

Fig. 5 Relation between classes

querying for a tutor and providing the most suitable options from the following classes in the obtained ranking of the current learners. With this approach, the tutor retrieval becomes a matter of adequately specifying querying parameters. The proposed mechanism offers the possibility of obtaining any feasible solutions, somewhere between the next learner who resides in the class with the following knowledge weight value up to the top-class learners in ranking. From this perspective, several parameters are defined, among which one manages the proximity of the class from which tutors are retrieved. This method is accomplished by setting an integer value called NEXT, which defines how the procedure will look for tutors. For example, if next is set to value "1", the tutors will be retrieved from the next class while is next is set to value MAX then tutors will be retrieved from the best class in ranking. Figure 5 presents an intuition regarding the relation between classes.

The current approach is set manually, but automatically finding the personalised optimal value needs further analysis and experiments. Another parameter regards the number of retrieved tutors currently, this value is set manually, and thus the most similar learners from the target class are returned as results set. Another parameter tuning is accomplished by considering two types of feature subsets: one regards assessed knowledge weight and one regards messaging activity. Ideally, a learner looking for a tutor may want access to colleagues with large values in knowledge weight that are also very active in messaging activity. Still, it may not be the case for practical reasons that such learners actively engage in tutoring colleagues with much knowledge level or messaging activity. Therefore, a trade-off between this feature and a combined utility function may be needed to maximise genuine cooperation between querying learners and recommended tutors. The current approach allows querying learners to set up the values for any of the introduced parameters and therefore obtain a set of prospective tutors.

4 Conclusions

This chapter presents a roadmap for performing interface optimisation for Tesys e-Learning system. The designed questionnaires and surveys provide valuable feedback from students and professors regarding various functionalities, such as communication or testing. The task of the proposed analysis is to reveal the factors that influence the perceived ease of use.

The usability evaluation of an e-Learning platform may be performed after the business logic for the functionalities (i.e., communication, testing, etc.) has been implemented and is ready to use. As some of the functionalities are classical and regard classical features, some features rely on complex data analysis pipelines that integrate machine learning algorithms. From this perspective, this chapter presents a tutor recommender system that relies on student's classification, which uses a classical decision tree algorithm for prototype application. The attempt to increase students' engagement by providing peer tutors that are correctly classified by the machine learning pipeline represents a prototype system that may be further improved in various ways. One line of research regards improving the classifiers in terms of their accuracy: their ability to label students correctly. Still, besides this data analytics perspective, evaluating the relevance of recommended tutors represents a challenging task.

We conclude that usability evaluation needs to be performed from two perspectives: the interface layout and workflow, and the effectiveness of advanced functionalities (i.e., machine learning based recommender systems) implemented within the platform.

References

1. Mostow, J.B.: An educational data mining tool to browse tutor-student interactions: time will tell. In: Proceedings of the Workshop on Educational Data Mining, National Conference on Artificial Intelligence, pp. 15–22. AAAI Press (2005)
2. García, E.R.: Drawbacks and solutions of applying association rule mining in learning management systems. In: Proceedings of the International Workshop on Applying Data Mining in e-Learning (ADML) (2007)
3. Popescu, P.S., Mihaescu, M.C., Mocanu, M., Ionascu, C.: Evaluation of the tesys e-learning platform's interface. In: RoCHI, pp. 86–90 (2016)
4. Mihaescu, M.C., Popescu, P.S., Ionascu, C.M.: Questionnaire analysis for improvement of student's interaction in tesys e-learning platform. Roman. J. Hum. Comput. Inter. **10**(1) (2017)
5. Park, S.Y.: An analysis of the technology acceptance model in understanding university students' behavioural intention to use e-learning. Educ. Technol. Soc. **12**(3), 150–162 (2009)
6. Popescu, P.S., Mihaescu, C., Mocanu, M., Ionascu, C.: Exploring the perceived ease of use by professors in tesys e-learning platform. In: RoCHI—International Conference on Human-Computer Interaction, pp. 15–20 (2017)
7. Ardito, C.C.: An approach to usability evaluation of e-learning applications. Univ. Access Inf. Soc. 270–283 (2006)
8. Costabile, M.F.: On the usability evaluation of e-learning applications. In: Proceedings of the 38th Annual Hawaii International Conference on System Sciences. Big Island, HI, USA (2005)

9. Liaw, S.S.: Investigating students' perceived satisfaction, behavioral intention, and effectiveness of e-learning: a case study of the Blackboard system. Comput. Educ. **51**(2), 864–873 (2008). https://doi.org/10.1016/j.compedu.2007.09.005
10. Bowman, D.A.: A survey of usability evaluation in virtual environments: classification and comparison of methods. Teleoper. Virtual Environ. **11**(4), 404–424 (2002)
11. Ernst, C.P.: Students' acceptance of e-learning technologies: combining the technology acceptance model with the didactic circle (2014)
12. Nielsen, J.A.: Heuristic evaluation of user interfaces. In: Proceedings of the SIGCHI Conference on Human Factors in Computing Systems, pp. 249–256. ACM (1990)
13. Weizhe Liu, L.K.: Semiautomatic annotation of MOOC forum posts. In: Chapter State-of-the-Art and Future Directions of Smart Learning Part of the series Lecture Notes in Educational Technology, pp. 399–408 (2016)
14. Romero, C.E.: Web usage mining for predicting final marks of students that use Moodle courses. Comput. Appl. Eng. Educ. 135–146 (2013)
15. Burdescu, D.D., Mihaescu, M.C.: TESYS: e-learning application built on a web platform. In: ICE-B, pp. 315–318 (2006)
16. Mihaescu, M.C., Popescu, P.S., Ionascu, C.M.: Intelligent tutor recommender system for online educational environments. In: EDM, pp. 516–519 (2015)

Developing New Algorithms that Suite Specific Application Requirements

M. C. Mihăescu, P. S. Popescu, and M. L. Mocanu

Abstract This chapter describes a newly developed algorithm for ranking by classification. We present the structure, class diagram and pseudocode of the main implemented specific functions. In terms of the implementation usage, we present an advanced classifier visualization panel and a sample usage for determining tutors. The implementation is provided as an open-source official Weka package. The proposed approach and roadmap opens the way for further development in adding new functionalities to the existing package, integrating, and using the package as client software for other systems or even taking the same approach to develop different new algorithms.

Keywords Developing a new algorithm · Ranking by classification · Developing an official Weka package

1 Introduction

When developing an e-Learning system, specific application requirements may not be tackled directly by using machine learning algorithms or even complex data analysis workflows. This situation needs particular attention because custom algorithmic approaches need to be designed in an efficient way such that the addressed problem gets a tractable solution. The condition usually reduces to the fact that the used package or library does not contain an algorithm the implements the behaviour needed by the application. Therefore it calls for new algorithmic design and implementation. Once this decision has been made, the software architect has two choices: (1) to design and implement the needed algorithm specifically for the tackled problem and platform, or (2) to isolate the logic of the algorithm into a stand-alone package that will further be used within the application, and potentially be used by other researchers within their developments.

M. C. Mihăescu (✉) · P. S. Popescu · M. L. Mocanu
Department of Computer Science and Information Technology, University of Craiova, Str. A.I. Cuza, Nr. 13, Craiova, Romania
e-mail: cristian.mihaescu@edu.ucv.ro

© The Author(s), under exclusive license to Springer Nature Switzerland AG 2022 147
M. Mihaescu and M. Mihaescu (eds.), *Data Analytics in e-Learning: Approaches and Applications*, Intelligent Systems Reference Library 220,
https://doi.org/10.1007/978-3-030-96644-7_8

Of course, the first approach represents the easy way and is usually followed by proprietary software. The second approach is more targeted for open-source software, as the new package may be developed within an open-source library and make it publicly available for the entire community of developers.

This chapter addresses the problem of developing a new classifier within an open-source library. We present the design of the classifier from an architectural perspective, the implementation, a visualization plugin, and a sample usage in the context of student search and the case of tutor's retrieval.

Development of the classifier as a package requires that all constraints of an open-source library are met. The package manager of the core application may download and deploy the new algorithm within the context of the developed application.

This chapter further presents the design of the classifier, its general architecture, the details of implementation, and a visualization plugin that may perform specific advanced algorithmic functions [1].

Many classification algorithms have been previously implemented in WEKA, but none of them can provide a data model that is overloaded with instances. Considering the previous observation, there are no WEKA visualization methods that can display the data in the model in an efficient way. There are no available parsing methods ready to develop such functionalities. Parsing leaves is another task that is missing in WEKA. It is crucial because instances from neighbour leaves have high similarity and share many features with similar values.

A real-life scenario in a training dataset can have many features that describe the instances. A data analyst must examine a decision tree, understand the rule that derives from an explicit decision, and then provide accurate conclusions. This chapter will tackle classification and visualization issues by adding new functionalities and improving the decision tree visualization.

2 Related Work

These days large amounts of data can be obtained from several research areas or applications used in the industry. Data mining and knowledge extraction [2] from raw data becomes more and more critical. Analysts gather many variables/features from this vast amount of data, analysts gather many variables/features, and many machine learning techniques are needed to face this condition. There are many application areas such as medical, economics (i.e., marketing, finances, or others), software engineering or, in our case, online educational research area [3] in which machine learning methods can be used. Educational data mining is a rising domain [4] in which a significant amount of work has been done. Because the application domains are rising continuously, the tools that support the machine learning methods must live up to the market standards providing good enough performances and intuitive visualization techniques. Currently, many tools implement many machine learning algorithms in various programming languages such as Java, Python, R, etc. Tools like KEEL [5], RapidMiner [6], WEKA [7], Mahout [8] or Knime [9] represent only a few examples from the available ones from which here [10] there are the most popular ones. RapidMiner is a visual analytics system, previously known as YALE, which delivers an integrated environment for machine learning, data mining, and predictive data analytics. Another tool is Keel, a software package of machine learning tools focused on evaluating evolutionary algorithms. Further, KNIME (Konstanz Information Miner) is a flexible data investigation platform, provided as a plugin for Eclipse, which offers a graphical workbench and numerous mechanisms for data mining and machine learning. Implemented in Java, Mahout is a very scalable machine learning library built on the Hadoop framework, which implements the MapReduce programming model. This model supports distributed processing [11] of large sets of data across groups of CPUs.

We choose to use WEKA to integrate the developed algorithm and run experiments because it is one of the most comprehensive data mining and machine learning libraries. Moreover, WEKA is a favourite tool used in numerous research areas, generally adopted by most EDM societies. We mention that Weka benefits from constant development and improvement.

WEKA is implemented in Java and encapsulates an extensive collection of algorithms that addresses many data mining and machine learning tasks. We mention the most used classes of algorithms: pre-processing, clustering, regression, classification, association rules and neural networks.

In some cases, these machine learning algorithms refer only to the underlying implementation. One particular feature that WEKA possesses is that it has an implemented package manager, which simplifies the developer's contribution process. Thus, through the package manager WEKA may use both official and unofficial packages. This approach is beneficial because if an algorithm that can solve your problem is available in a package, you can add that specific package to the project and use it accordingly along with the WEKA core. Furthermore, you don't need to be a proficient software developer to do that; you do not necessarily need to develop code because you install the package and then use the algorithm from the Weka software interface like the ones included in the core distribution.

According to real-life experiences, many of the included algorithms can barely be used because of their lack of flexibility. For example, in a standard implementation of decision trees from WEKA, we can perform a classification task, but we cannot access a particular instance from a specific leaf. Let us consider we have a training data set, and we generate the tree model. When we want to locate the example "X" in the decision tree, we cannot do that in the application interface, nor when you add the WEKA library in your code. This situation is a disadvantage because retrieving the instance leaf sometimes provides more information than retrieving its class. Usually, when performing a classification task, the data analyst divides test instances into categories that have little meaning from the application domain of perspective.

One aspect that distinguishes WEKA from other comparable software is its architecture that allows developers to contribute practically. All the implementation work that needs to be done refers to making a specific folders arrangement, adding the "*.jar*" file to the archive, completing a "*description.props*" file, and then the build script.

WEKA is an open-source machine learning library that allows developers and researchers to contribute very quickly. Over twenty years since WEKA had its first release [12], regular contributions were added. Not only machine learning algorithms were developed; for example, in 2005, a text data mining module was implemented [13]. A summary of the available applications is presented in [7]. Several classifiers were implemented and contributed as packages to WEKA. Still, one exciting classifier was built in 2007 based on a set of sub-samples [14] and related to C4.5 [15], which have its implementation called J48 [16] in WEKA. Other classifiers refer to voted decision trees and voted decision stumps or the "Alternating Decision Trees Learning Algorithms" (Mason), which simplifies the regular decision trees. This type of classifier is relatively easy to interpret, and the rules are usually smaller in size. Regular decision trees, such as C4.5 algorithm, were expanding nodes in a depth-first order; an upgrading came from algorithms called "Best-first decision trees" [17], which grow nodes in a best-first order and a package with these trees was contributed to WEKA.

Some other type of contributions refers libraries of algorithms that can be integrated into WEKA. One of the first libraries contributed and is JCLEC [18], an evolutionary computation framework that has been successfully employed for developing several evolutionary algorithms. Another environment for machine learning and data mining knowledge discovery that was contributed to WEKA is R [19]. This contribution was designed to include different sets of tools available in a single combined system from both environments. There is also a package that provides WEKA's core algorithms in the R programming language.

Also, as related work, we must consider some of the development of the C4.5 algorithm. A few years ago, a new fast decision tree algorithm [20] was implemented and based on the experiments provided by the authors. The classifier outperforms even C5.0, which is the commercial and more scalable implementation of C4.5.

3 Building a New Classifier

The package that implements the new classifier is designed to be used by researchers using the WEKA Explorer and developers in their Java applications. Initially, the package implementing the new classifier was under development, offering more functionalities as a *.jar* file for developers than in the explorer interface of WEKA. Later, the package becames official with full integration, allowing users to access it from the WEKA's interface and developers as *.jar* file.

Figure 1 presents the design of the algorithm and what is its relation with the WEKA core. In the upper part of the figure, the classifier is depicted divided into two main modules: visualization and the algorithm itself. As we can see at the next level, both modules are further divided. All the functionalities can be installed in WEKA via the package manager option. Then, using the explorer, we can do data analysis tasks using a model loaded with instances and its associated visualization techniques.

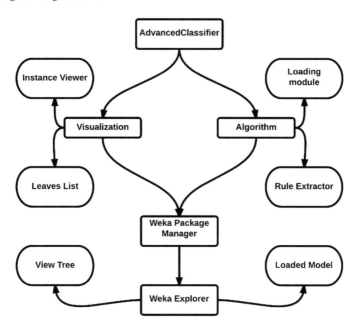

Fig. 1 Package overview diagram

3.1 The General Architecture of the Classifier

The package is archived as a *.zip* archive and is structured according to the WEKA rules. The archive unpacks to the existing directory. It contains a build script, the source files, a folder with the required libraries, a property file required by WEKA for installing and handling the package, and the most important thing: the *.jar* file. A comprehensive structure of the package and its organization is presented below in Fig. 2.

Figure 3 presents the package's class diagram. This diagram contains all the java files from the project along with their relations. As we can see in Fig. 3, we have two types of classes: composed and independent. Independent classes are part of the model section of the Model-View-Controller (MVC) architecture or classes that perform one-time tasks like "*WekaTextFileToXMLTextFile*", which generates an XML file based on the text-based file returned by WEKA. The classes that are composed have specific dependencies, and their relations are joint across packages. One critical class is *AdvancedClassifierTreeLeaf.java* which serializes the leaves of the decision tree along with rules (or path) from the root to a specific leaf.

Fig. 2 Structure of the package

```
<current directory>
 +-AdvancedClassifier.jar
 +-Description.props
 +-build_package.xml
 +-src
 |  +-main
 |     +-java
 |        +-resources
 |        |  +-background_node.png
 |        |  +-background_leaf.png
 |        |  +-background_leaf_pressed.png
 |        |  +-font_node.ttf
 |        |  +-font_decision.ttf
 |        +-weka
 |           +-classifiers
 |           |  +-trees
 |           |     +-model
 |           |     |  +-AdvancedClassifierTree.java
 |           |     |  +-AdvancedClassifierTreeBaseNode.java
 |           |     |  +-AdvancedClassifierTreeNode.java
 |           |     |  +-AdvancedClassifierTreeLeaf.java
 |           |     |  +-BaseAttributeValidator.java
 |           |     |  +-NominalAttributeValidator.java
 |           |     |  +-NumericAttributeValidator.java
 |           |     |  +-Constants.java
 |           |     +-AdvancedClassifier.java
 |           |     +-WekaTextfileToXMLTextfile.java
 |           +-gui
 |              +-visualize
 |                 +-plugins
 |                    +-AdvancedClassifierTree.java
 |                    +-AdvancedClassifierTreePanel.java
 |                    +-BaseNodeView.java
 |                    +-AdvancedClassifierTreeNodeView.java
 |                    +-AdvancedClassifierTreeLeafView.java
 |                    +-ConnectingLineView.java
 +-lib
    +-weka.jar
    +-simple-xml.jar
    +-rt.jar
```

3.2 Implementation of the Classification Algorithm

The algorithm aims to generate specific rules (depending on what training dataset we use) for every decision tree leaf. These rules are developed by starting from the tree's root and parsing the splits to the specified leaf. Therefore, each decision path leading to a leaf is interpreted into a rule that includes the name of the feature and the value on which the split in the decision tree was performed. For each feature defined by WEKA, we will have a specific rule that matches the corresponding feature.

For this purpose, we defined an abstract class to act as a base class for any of the customs rules. The class is called *BaseAttributeValidator* and defines the required functions that a superclass has to extend: a *clone()* function required by the workflow of the whole system and methods that validate an instance or group of instances have the needed values of the feature targeted by the rule. Currently, the only implemented rules can handle two feature types: "NOMINAL" and "NUMERIC".

The function that implements the rule that validates the simple type features is called *NominalAttributeValidator()* and takes the name of the targeted feature and a string variable representing the accepted value of the feature. It may be used to

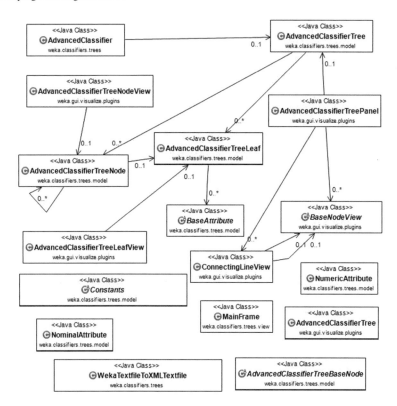

Fig. 3 Classifier's class diagram

handle the numeric attributes by calling the *NumericAttributeValidator() function*. It also receives the name of the feature and either a specific value or the boundaries of an interval.

Further, we present a short overview of the algorithm for which we chose a straightforward approach. In the first step, the algorithm takes instances from a provided "*.arff*" file using the functions provided and implemented by WEKA. The second step calls the desired classification algorithm on them. Currently, the only supported classification algorithm is J48, but extending to other decision tree classifiers is considered future work.

Using the text representation of the generated model and a predefined set of rules and tags, an XML is then generated. This is critical during the whole workflow because the structured XML format lets us store the base model for our decision tree. The deserialization of the object is performed using a third-party Java library (i.e., *Simple XML*). The model built this way contains a list of leaves and nodes with the following meaning: each node matches a decision made in the tree model; the information stored in each object (node) refers the data about the name of the actual feature, operator, and the value on which was made the split; and the results obtained after making the decision (the output leaf or a list of the successor nodes).

We determine the set of rules by using this approach and the set of features provided by WEKA. This step is achieved by traversing the tree model from the root to the leaf and iteratively composing the rules that define each leaf. The algorithm's configuration is finally ended by loading the training dataset into the model and producing a **dense tree**.

Once the dense tree (i.e., the tree with loaded instances) is available within the classification model, the data can be straightforwardly handled for running specific tasks. The currently developed method includes basic handling of instances per leaf, i.e., loading new instances into the tree model and obtaining the part of the dataset in each leaf and computing predecessor and successor.

3.3 Visualization Plugin

We have designed a custom panel for the visualization plugin that displays the decision tree and show all the data available in the leaves. The panel's constructor requires the decision tree model as a parameter and adds the corresponding view layouts to the interface. We developed a specialized class that uses WEKA's *TreeVisualizePlugin* interface. After adding the package through the WEKA's Package Manager and selecting this visualization choice, a new JFrame is created for accomodating the custom panel.

Figure 4 presents a dataset sample. Several validation tests have been performed to validate the classifier and its extra functionalities, but we used only three attributes and 788 instances for this case study. The feature called "userid" does not provide any information for instances, but it can merely be used for instances localization into leaves.

Figure 5 presents a screenshot of the tree generated based on the dataset from Fig. 4. Every node contains the feature's name, and each decision is printed on top of

Fig. 4 Sample fragment from the dataset

```
@RELATION StudentClass

@ATTRIBUTE userid numeric
@ATTRIBUTE nrHours numeric
@ATTRIBUTE avgMark numeric
@ATTRIBUTE present {NO,YES}
@ATTRIBUTE class {low,high}

@DATA
4,17,6.025,NO,low
5,11,7.0,YES,low
7,30,7.375,NO,low
8,10,7.714286,NO,low
9,43,5.1666665,YES,high
10,31,4.5,YES,high
```

Fig. 5 Tree sample

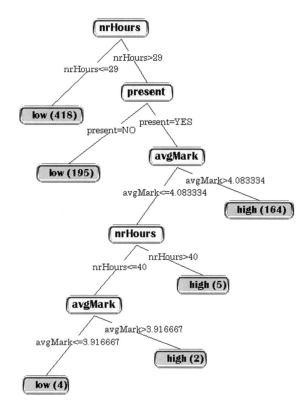

the connecting line. Undoubtedly, each leaf can be clicked, and the set of enclosed instances is displayed. At this moment, all the means to visualize and handle the predecessors/successors, outliers and other relevant information are developed and integrated into the official package.

Regarding implementing specific functionalities encapsulated in the advanced classifier, we present some of the implemented procedures and details below. The workflow is also described, and the whole mechanism brings new functionalities for the users. One immediate thing that needs to be mentioned is that the data loading module created new data analysis opportunities for researchers. As future work, we want to implement different other types of attributes supported by WEKA like "String", "DATE", or "Relational."

A sample of the algorithm for computing the predecessor or successor of a specific leaf is presented below.

createLeafList(node) returns a vector containing the leafs {
    stackOfParents := leafList := empty stack

    while (not stackOfParents.isEmpty() or node != null){
        if (node != null)
            stackOfParents.push(node)
            node := node.left
        else
            node := stackOfParents.pop()
        if (node.type == leaf)
            leafList. add(node)
        node: = node.right
    }// end while
    return leafList
}

computeSuccessor(leafList, node, N) returns successor of node in leafList {
    successorNode := NULL
    while (leafList.hasMoreElements()){
        if (node = = leafList.next())
            for (i=0; i<N; i++)
                if (leafList.hasMoreElements())
                    successorNode = leafList.next()
                else
                    return NULL
    }// end while
    return successorNode
}

computePredecessor(leafList, node, N) returns predecessor of node in leafList {
    predecessorNode := null
    while (leafList.hasMoreElements()){
        if (node = = leafList.next())
            for (i=0; i<N; i++)
                if (leafList.hasMoreElements())
                    successorNode = leafList.previous()
                else
                    return NULL
    }// end while
    return predecessorNode
}

As we can see in the algorithms presented above, three methods are used to compute the predecessor, the successor of a specified node and to create the list of leaves. The first method is a generic one and must be called before any of the following two methods for building the list of leaves by parsing the tree from left to right. This step eliminates unnecessary operations needed later if we compute the list in every other function. Due to the in-order parsing of the tree, we will get an ordered list of leaves, simplifying the computation of successor and predecessor.

Function *computeSuccessor()* is being called with three parameters: **leftist**—which is returned from *createLeafList* and contains all the leaves in the tree well-ordered; **Node**—the node for which we want to find a successor and **no**—the successor's grade. For example, computing the direct successor requires value 1 for the successor's degree, and if we would like to find the 4th successor, the value of no shall be 4. As we can see in the above-presented pseudocode, this function will return the exact successor of the desired node.

3.4 Demo of the New Classifier

After installing the package into Weka and loading the dataset, we can train a classifier (i.e., J48) and build a model. After this step, we can run the plugin by right-clicking on the available model and selecting from visualization the advanced classifier.

Figure 6 presents an example of a model built on a custom-made dataset with a small number of attributes to validate. On the right panel, we have the available functionalities which can be called on the already trained model.

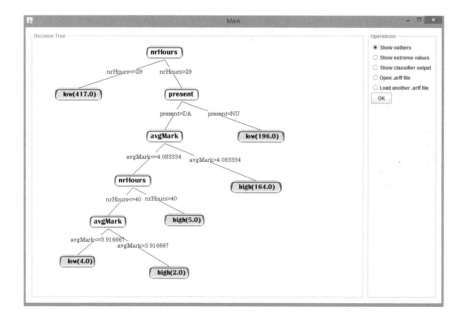

Fig. 6 Advanced classifier visualization panel

We used a synthetic dataset to build this model because we need a controlled environment that can produce an interpretable and accessible to validate the model. Below we have a snip of the dataset along with the validation metrics.

As we can see in Fig. 7a, the dataset has a large number of instances (788) but a low number of attributes (3) and the accuracy obtained using the J48 algorithm is over 99%. The tree has few splits and few internal nodes and leaves, allowing us to read the model and quickly analyze the obtained model.

Figure 8 presents the result after selecting a leaf from the tree once the model was trained and loaded with data. In the first screenshot, we can see data regarding instance 49 from the leaf, and we have two options: visualize the successors and predecessor. For each of these options, we can select the level of the successors and predecessors; this refers to the distance, counting the number of leaves, from the

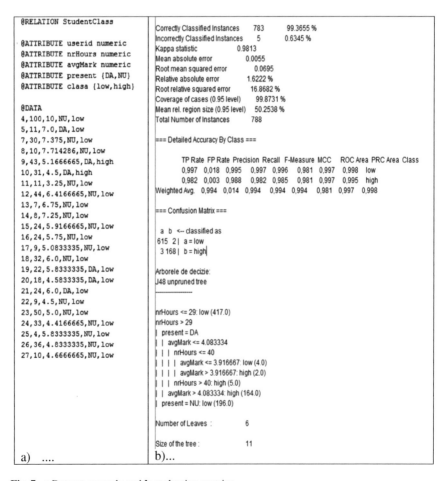

Fig. 7 **a** Dataset example and **b** evaluation metrics

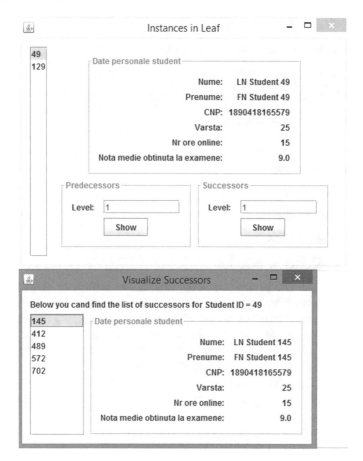

Fig. 8 Visualizing a leaf

selected node to the one we desire to analyze. In the second screenshot, we have the level, one successor, for instance, 40 and our information regarding them.

3.5 Sample Application: Determining Tutors Using the New Classifier

The primary input of the server application is the active repository. This raw data gathered from the database is converted to a *.arff* file, used to train the classifier [21]. Below is presented a sample of a generated *.arff* file.

@ATTRIBUTE userid numeric

@ATTRIBUTE nrHours numeric

@ATTRIBUTE avgMark numeric

@ATTRIBUTE messagingActivity {NO, YES}

@ATTRIBUTE noOfTests numeric

@ATTRIBUTE avgMessageLenght {short, long}

@ATTRIBUTE class {low, high}

@DATA

1, 29, 6.025, NO, 4, long, low

2, 14, 5.1666665, YES, 6, long, low

3, 51, 4.5, YES, 4, long, high

4, 18, 6.75, YES, 2, short, low

5, 51, 6.4166665, NO, 6, long, high

In Fig. 9, we present the meaning of the attributes from the *.arff* file. We have the attribute's name in the left column and the right column its description.

Figure 10 presents the obtained decision tree from the training dataset. Parsing the leaves from left to right, we can find leaves containing lists of possible tutors. We implemented several functionalities for parsing and loading a decision tree model along with instances. Due to the dataset's structure, the decision tree contains ordered leaves. This approach can effectively find specific tutors that have a particular distance from the querying student. For example, if the student does not want

userid	The identification number of the student
nrHours	The total number of hours spent on the platform
avgMark	The average mark obtained at a discipline
messagingActivity	This feature represents a discretization of the on-line involvement regarding sent and received messages. A messaging activity greater than 50% of the average number of sent and received messages means YES and smaller means NO.
noOfTests	The total number of tests taken on the platform
avgMessageLength	This feature represents a discretization of the average message length. An average length greater than 160 characters means the student sends LONG messages and smaller means SHORT messages. This value has been chosen with respect to the limitation of the SMS message.
class	The class that the user belongs to, which can be low or high. Low/high values were assigned based on the final result obtained by the student.

Fig. 9 Attributes meaning

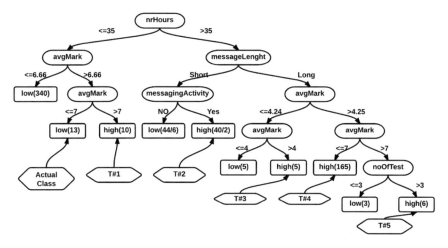

Fig. 10 Example of the obtained decision tree

iterative progress, he can use this feature to obtain tutors who are "*i* "steps ahead of him.

We further generate the XML version of the decision tree after creating the model in the memory. Here is the XML version of the decision tree. This XML version of the tree is sent to the client for visualization purposes. Below is a short example of the XML mentioned above.

```xml
<?xml version="1.0" encoding="UTF-8"?>
<DecisionTree type="J48Tree">
|      <Node attribute="nrHours" operator="&lt;=" value="35">
|      |        <Node attribute="avgMark" operator="&lt;=" value="6.666667">
|      |        |        <Output decision="low" info="(340.0)"/>
|      |        </Node>
|      |        <Node attribute="avgMark" operator=">" value="6.666667">
|      |        |        <Node attribute="avgMark" operator="&lt;=" value="7">
|      |        |        |        <Output decision="low" info="(13.0)"/>
|      |        |        </Node>
|      |        |        <Node attribute="avgMark" operator=">" value="7">
|      |        |        |        <Output decision="high" info="(10.0)"/>
|      |        |        </Node>
|      |        </Node>
|      </Node>
..............
        </Node>
</DecisionTree>
```

Below is the pseudocode for the function used to locate the classified student's actual and target class. This recursive function takes two parameters: a node element containing information from the XML and a boolean variable, stating whether the parent has been marked or not. A node is marked when the learner meets the requirement displayed inside the node (for example, if the "*messageLenght*" is "*LONG*"). The parent node will finally contain an additional attribute for the leaf that represents the student's actual class.

The function *findTutor* is called with root *node* as argument. The *isInPath* parameter is used for marking the nodes on the path from root to the actual class of the student. After traversal of the tree the node will contain the attributes and their values from nodes that were visited.

The return value is *targetNode* and is the closest leaf labeled as "high" and will contain tutors.

```
*/
findTutor (Student actualStudent, Node node, boolean isInPath) returns targetNode{
    childrenList := list of children of current node
    FOR each node in childrenList
        IF (node is internal node && node(atributeValue) maches actualStudent){
            #add (attributeName, attributeValue) to the node
            node(isInPath) := true
            findTutor(node, false);// recursive call for child node
        }ELSE{// node is leaf
            IF (class attribute is "high")
                targetNode := current node;// we have finished the traversal
of the tree
                return targetNode
        }
    }//end for
}
```

The client application automatically determines the class (i.e., label) of the student interested in finding a tutor. The label represents the actual class of the student. In Fig. 11, the student's current class is marked with red, and the target class he wants to achieve is marked with green.

The inspection of the green node reveals to the student a list of classmates that may help him with knowledge as tutors. The student has the built-in messaging system at his disposal to contact his recommended tutors to find answers from the right colleagues.

Considering the aspects presented, we can perform a sample usage scenario for a scholar who seeks an appropriate tutor. Let us consider scholar *S1*, which has previously used the e-Learning platform and there is enough data logged about him and the following values for the features are recorded: *avgMark = 6.80, nrHours = 30, messagingActivity = No, noOfTests = 2, avgMessageLength = SHORT*. After running the program, he will find out that the second leaf is his actual one when

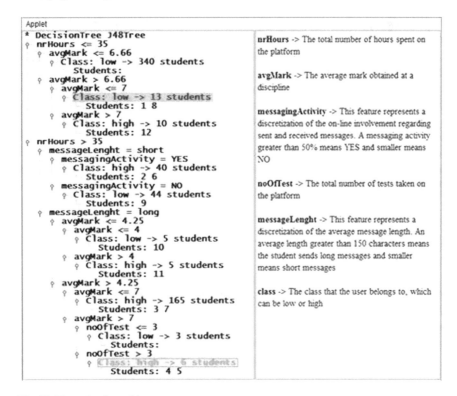

Fig. 11 Example of matching tutors

parsing the tree from left to right, so he must work hard to catch up with his colleagues. Assuming he wishes to spend the time gradually for learning from his colleagues, the application can recommend him colleagues from the third leaf of the tree model (1st level successors). In this case, he will receive the contact details of his classmates from that specific leaf. After improving his learning performance and becoming part of the target leaf, he will go further to the next step. Suppose the actual student, wants to talk to the best student directly (in one step). In that particular case, the system will give him sufficient contact details to get in touch with the greatest tutors available (students that belong to the green leaf from Fig. 11).

Two there are two arguments that needs to be taken into consideration when evaluating the system's performance and determining if it is performing well enough or should be improved. The first arguments is that the learners from a green leaf are always positioned further to the right (to the bottom in Fig. 11) within the tree than the students from a leaf coloured in red. This observation respects the specified dataset because in our decisions tree, on the right side, we have more hours spent, a greater mark, a bigger number of tests. This observation means that the students positioned further to the right have better performance and are more active, so traversing the tree from left to right will always find better learners. The second argument is that they belong to a high class, so their results now are classified by the algorithm as "high", and therefore they can implicitly be a good example for the other students.

One relevant question could be "why we do not choose the students who have the *avgMark* greater than 7, a high class, considering those students being better and closer to the actual students?" The reason is that if we aim to improve only one attribute of a student, the chance of failure is bigger. Computing the successor of a specific leaf will not constantly give us the best learners to be recommended. If we parse the tree, we can see that our learners have an *avgMark* better than 7, which was the reason for the actual learner being classified as low, but they also have better attributes values for hours spent on learning or the number of tests used for training to get prepared for the exams.

4 Conclusions

This chapter presented a complete flow for developing a new machine learning algorithm within an open-source Weka package and its integration within an online learning environment. The development and design of the algorithm should consider some critical aspects. First, it should respect the regulations imposed by the open-source package for the library which it is developed. This aspect decreases to several factors like usage of main data structures and algorithms already included into the opensource package, respect a proper architectural design which has a well isolated implementation from the package and particularly from the e-Learning system in which it will be integrated. Properly referring these aspects will allow further integration within other online learning environments or other tools that need the specifics of the newly designed and implemented machine learning algorithm.

Second, the new algorithm needs to provide an implementation whose design is well isolated from the package within which it is implemented and from the client application. This method will allow further extension of the new machine learning algorithm and adjustment to other specific functionalities further required or in other applications.

In this chapter, after presenting a list of well-known packages that implement different algorithms, we present the motivation for choosing Weka machine learning library as a package tool. Section three offers the structure of the package, its class diagram and the pseudocode of specific algorithms that were implemented in the *RankerByDTClassification* official package. As an specific application of the package, we present an advanced classifier visualization panel and a sample usage for establishing tutors using the new classifier.

Code Availability

Implementation of the new algorithm is part of the Weka packages official list [22] and is provided as open-source software that is freely available [23].

References

1. Popescu, P.S., Mocanu, M., Mihaescu, M.C.: Integrating an advanced classifier in WEKA. In: EDM (Workshops) (2015)
2. Fayyad, U., Piatetsky-Shapiro, G.: The kdd process for extracting useful knowledge from volumes of data. Commun. ACM **39**(11), 27–34 (1996)
3. Romero, C., Ventura, S.: Educational data mining: a survey from 1995 to 2005. Expert Syst. Appl. **33**(1), 135–146 (2007)
4. Selwyn, N.: Data entry: towards the critical study of digital data and education. Learn. Media Technol. **40**(1), 64–82 (2015)
5. Alcalá-Fdez, J., Fernández, A., Luengo, J., Derrac, J., García, S., Sánchez, L., Herrera, F.: Keel data-mining software tool: data set repository, integration of algorithms and experimental analysis framework. J. Mult.-Valued Logic Soft Comput. **17** (2011)
6. Mierswa, I., Wurst, M., Klinkenberg, R., Scholz, M., Euler, T.: Yale: Rapid prototyping for complex data mining tasks. In: Proceedings of the 12th ACM SIGKDD International Conference on Knowledge Discovery and Data Mining, pp. 935–940 (2006)
7. Hall, M., Frank, E., Holmes, G., Pfahringer, B., Reutemann, P., Witten, I.H.: The WEKA data mining software: an update. ACM SIGKDD Explor. Newsl. **11**(1), 10–18 (2009)
8. Owen, S., Friedman, B.E., Anil, R., Dunning, T.: Mahout in Action. Simon and Schuster (2011)
9. Berthold, M.R., Cebron, N., Dill, F., Gabriel, T.R., Kötter, T., Meinl, T., Wiswedel, B.: KNIME-the Konstanz information miner: version 2.0 and beyond. AcM SIGKDD Explor. Newsl. **11**(1), 26–31 (2009)
10. Most Popular Machine Learning Software Tools: https://www.softwaretestinghelp.com/machine-learning-tools/ (2021)
11. Shvachko, K., Kuang, H., Radia, S., Chansler, R.: The hadoop distributed file system. In: 2010 IEEE 26th Symposium on Mass Storage Systems and Technologies (MSST), pp. 1–10. IEEE (2010)
12. Holmes, G., Donkin, A., Witten, I.H.: Weka: a machine learning workbench. In: Proceedings of ANZIIS'94-Australian New Zealnd Intelligent Information Systems Conference, pp. 357–361. IEEE (1994)
13. Witten, I.H., Frank, E., Trigg, L.E., Hall, M.A., Holmes, G., Cunningham, S.J.: Weka: Practical machine learning tools and techniques with Java implementations (1999)
14. Pérez, J.M., Muguerza, J., Arbelaitz, O., Gurrutxaga, I., Martín, J.I.: Combining multiple class distribution modified subsamples in a single tree. Pattern Recogn. Lett. **28**(4), 414–422 (2007)
15. Quinlan, J.R.: C4. 5: programs for Machine Learning. Elsevier (2014)
16. Loh, W.Y.: Classification and regression tree methods. Encyclopedia Stat. Qual. Reliab. **1**, 315–323 (2008)
17. Shi, H.: Best-First Decision Tree Learning. Doctoral dissertation. The University of Waikato (2007)
18. Ventura, S., Romero, C., Zafra, A., Delgado, J.A., Hervás, C.: JCLEC: a Java framework for evolutionary computation. Soft. Comput. **12**(4), 381–392 (2008)
19. Hornik, K., Buchta, C., Zeileis, A.: Open-source machine learning: R meets Weka. Comput. Stat. **24**(2), 225–232 (2009)
20. Purdilă, V., Pentiuc, ŞG.: Fast decision tree algorithm. Adv. Electr. Comput. Eng. **14**(1), 65–68 (2014)
21. Mihăescu, M.C., Popescu, P.Ş, Burdescu, D.D.: J48 list ranker based on advanced classifier decision tree induction. Int. J. Computat. Intell. Stud. **4**(3–4), 313–324 (2015)
22. Weka official package list: https://Weka.sourceforge.io/packageMetaData/
23. Weka link for the RankerByDTClassification official package: https://Weka.sourceforge.io/packageMetaData/RankerByDTClassification/index.html

Printed in the United States
by Baker & Taylor Publisher Services